For Douglas

Contents

Introduction

Until tissue components could be identified histochemically, biologists and pathologists had to remain content with the morphological assessment of their microscopical preparations. Isolated attempts to interpret structural changes in chemical terms were due largely to botanists and pioneered by the eminent Frenchman Raspail who, 150 years ago, used iodine to detect starch and first applied the xanthoproteic test to demonstrate proteins microscopically. Such methods were somewhat rudimentary and not until 1936, when Lison first published his *Histochemie animale*, was histochemistry really recognized as an established science. Since that time sophisticated techniques have been devised to identify and localize, or demonstrate the activity of, a wide range of compounds occurring either naturally or abnormally in plant and animal tissues. Recently some of these methods have been successfully developed for quantification or have been adapted for electron microscopy and immunocytochemistry. For a fascinating historical review of histochemistry and its evolution the reader is referred to Pearse (1968, pp 1–12).

Of all histochemical topics, that of lipids has, perhaps understandably, had limited appeal and evoked rather less enthusiasm than other subjects. Several reasons for this unpopularity can be suggested.

1. The relatively late appearance of techniques for lipid histochemistry, largely because the chemical structure of the lipids themselves had not long been defined;

2. Methods tend to be laborious, unreliable, and require 'special' reagents that are not always to hand;

3. Suitable control material may not be readily obtainable;

4. Lipid methods cannot be applied to routinely processed paraffin or resin sections and until comparatively recently, before the advent of immunohistochemistry, cryostat sections of fresh frozen material were not universally available;

5. The demand for lipid histochemistry is relatively infrequent.

However, on the rare occasions that lipid techniques are requested, the results may be of paramount importance diagnostically and consequently the histochemistry of lipids has been considered a worthwhile topic for this series of RMS handbooks.

The aims of this handbook are threefold:

1. To dispel inhibitions about lipids and to overcome any reluctance towards their demonstration histochemically;

2. To recommend a series of reliable and straightforward methods that are selective for individual lipids and within the scope of the general microscopist;

3. To show that the detective work entailed when such methods are applied to pathological and biological problems may be intellectually satisfying.

In order that these techniques can be employed to best advantage and their reactions interpreted confidently, it is essential that the microscopist have a knowledge of the nature of lipids themselves and of their structural and chemical properties which are to be exploited in histochemical methods, together with an awareness of the limitations of such methods. Since this is essentially a working handbook, the reader will be spared much detail but, for a more comprehensive account of all aspects of lipid histochemistry, the reader is referred to Pearse (1968, Chapter 12), Adams (1965, Chapter 2), and Bancroft and Stevens (1982, Chapter 12).

Structural, chemical, and physical properties of lipids

Table 1 attempts to classify the main lipid groups with which we shall be concerned and to establish the nomenclature to be used throughout. Trivial names will be retained when subgroups are indistinguishable microscopically. Certainly this does not claim to be a complete classification in biochemical terms; rather a guide to the properties of lipids that can be considered relevant to histochemistry. Terminology itself has been a controversial issue and confusion arises even over the terms 'lipid' and 'fat'.

Lipids have been defined as 'naturally occurring fat-like substances that are soluble in organic solvents but not in water'. However, not all lipids are 'fat-like', a notable exception being crystalline cholesterol with a melting point of 144°C, whilst some phospholipids tend to be water-soluble. Lovern (1955) more realistically defined lipids as 'actual or potential derivatives of fatty acids and their metabolites'.

The major lipid classes consist of fatty acids linked to alcohols such as glycerol by ester bonds or to bases by amide bonds. Some contain in addition organic bases, phosphoric acid, and sugars. Above all it is the wide variation in their constituent fatty acids that determines the heterogeneity within each major class and the diversity between species throughout the plant, insect, and animal kingdoms.

The term 'fat' is generally applied to triglycerides, the most abundant lipids of both animals and plants. Those that are liquid are called oils, such as cod-liver and olive oils. The terms, 'fat' and 'oil', have no structural significance; nevertheless, the physical state of a lipid is of paramount importance in histochemistry. The melting point of a lipid will determine its reaction with an important group of organotropic Sudan dyes and is inversely related to the chain length and degree of saturation of its constituent fatty acids. Lipids that have a melting point around body temperature may crystallize at staining temperature and thus fail to react in tissue sections as they would have done *in vivo*.

Perhaps the most useful distinction in histochemical terms is that between hydrophobic and hydrophilic lipids. The surface properties of a lipid molecule determine its predeliction for either organic solvents or aqueous reagents. Polar phosphoryl and basic configurations within phospholipids cause them to be hydrophilic — water-miscible — whereas glycerol and cholesterol esters, waxes, and cholesterol itself contain a preponderance of nonpolar groups which render them hydrophobic. Their high surface tension at a lipid–water interface makes them impermeable to aqueous reagents and those that are in the liquid state have a special affinity for the organotropic fat stains.

Table 1. *Classification of lipids (after Adams 1965; Lake 1980, personal communication)*

Class	Member	Structural characteristics	Features influencing histochemistry
Unconjugated	Fatty acids	Saturated or unsaturated	Melting point (and sudanophilia) depend on number of double bonds. Hydrophobic
	Cholesterol	Tetracyclic compound characterized by its side-chain at the C_{17} position in the 5-membered ring	Melting point 144°C .: not sudanophilic. Hydrophobic. Birefringent in polarized light.
Conjugated			
1. Ester lipids	Cholesterol esters.	Fatty acid esters of cholesterol	Melting point depends on degree of saturation of constituent fatty acids. Hydrophobic.
	Mono-, di-, and triglycerides	Fatty acid esters of glycerol	
	Waxes	Esters of the higher alcohols	
2. Phospholipids		All contain phosphoric acid, longchain fatty acids, polyhydric alcohols, and variable nitrogenous bases and/or sugars.	Highly polar phosphoryl and basic groups cause phospholipids to be hydrophilic.
a. Glycerol-based:	Lecithins	Contain choline. Ester bond.	
	Cephalins	Contain ethanolamine or serine. Ester bond	Basic
	Plasmalogens	Ester and ether bonds.	Basic. Amide bond resists alkaline hydrolysis.
b. Sphingosine-based:	Sphingomyelins	Contain choline. Amide bond.	Neutral.
	Cerebrosides	Contain hexoses. Amide bond.	Highly acidic.
	Sulphatides	Sulphate esters of cerebroside.	Acidic. Water-soluble unless protein-bound.
	Gangliosides	Contain hexoses, hexosamines, neuraminic (sialic) acids. Amide bond.	

Reproduced from Bancroft and Stevens (1982) by courtesy of the editors and publishers.

2.1. Simple lipids
Fatty acids

R—COOH

In the animal kingdom these are straight-chain fatty acids with a terminal carboxyl group; they may be saturated or contain a variable number of ethylene (unsaturated) bonds. The fatty acids found in plants and micro-organisms are usually much more complex. The wide variety of fatty acids available for the construction of complex lipids accounts for the diversity of lipid subspecies, which are, however, indistinguishable histochemically. The melting point and Sudanophilia of a fatty acid depends upon the number of unsaturated bonds it contains as well as the length of its carbon chain. Commonly occurring saturated fatty acids are stearic (MP 69°C) and palmitic (MP 64°C) with chain lengths of 18 and 16 carbons respectively. Unsaturated fatty acids include oleic (with one double bond and a melting point of 14°C) and linoleic acid (MP − 11°C and two double bonds), both containing 18 carbons. These lipids are hydrophobic and will be Sudanophilic if molten at staining temperature.

Cholesterol

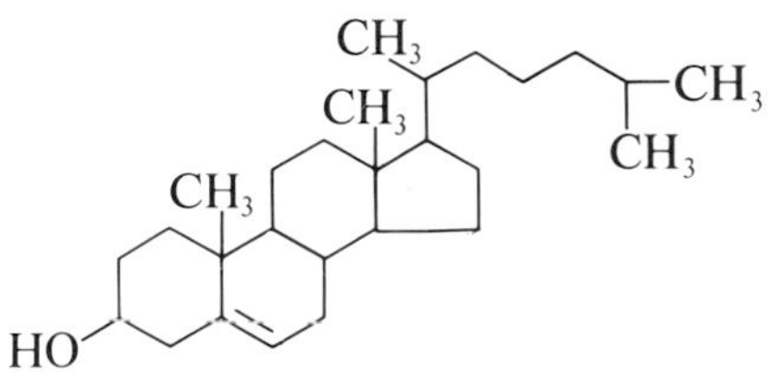

This steroid is a tetracyclic derivation of cyclopentenolphenanthrene, characterized by an aliphatic tail attached at the C_{17} position. In equimolar ratio with phospholipids, cholesterol constitutes the plasma membrane, aligning in a stable form so that the OH group is linked by hydrogen-bonding to the polar head group of a phospholipid molecule. Cholesterol is hydrophobic but crystalline (MP 144°C), therefore not Sudanophilic but will appear birefringent in polarized light.

2.2. Conjugated lipids
Esters

Cholesterol esters

Cholesterol can be esterified at the OH position by a variety of fatty acids, the identity of which will determine the melting point of the ester and its staining reactions, e.g. cholesterol oleate (MP 37°C) and cholesterol arachidonate (MP 23°C).

Cholesterol esters are hydrophobic and will be Sudanophilic unless crystalline at staining temperature.

Glycerol esters

$$CH_2-O-CO-R_1$$
$$|$$
$$CH_2-O-CO-R_2$$
$$|$$
$$CH_2-O-CO-R_3$$

These are esters of the trihydric alcohol glycerol with saturated and unsaturated fatty acids, oleic and palmitic occurring most commonly throughout nature. The ester molecules usually contains a mixture of different fatty acids. Triglycerides are the most abundant lipid in the higher animals and plants. They are hydrophobic and Sudanophilic unless crystalline.

Waxes

These are the esters of higher aliphatic alcohols and tend to be more inert than glycerol esters, having fatty-acid chain lengths of up to 30 carbons. Waxes are also hydrophobic and may be Sudanophilic if they are molten at staining temperature.

2.3. Phospholipids and glycolipids

These lipids contain long-chain fatty acids, polyhydric alcohols, phosphorus, various nitrogenous bases, and/or sugars. Phospholipids are amphipathic (liking both aqueous and organic solvents) with not only a hydrophilic polar 'head' on account of constituent phosphoryl and basic groups, but also a hydrophobic hydrocarbon 'tail'; hence their surfactant behaviour at oil—water interfaces. There are two main categories of phospholipid based on either *glycerol* or *sphingosine*. Glycolipids are those that additionally contain sugars. A variety of these glycolipids occur naturally in plants and glycosyl diglycerides are important in micro-organisms, but in the animal kingdom the glycolipids we are solely concerned with are cerebrosides and gangliosides.

Phosphoglycerides

Phosphatidyl esters — the lecithins and cephalins

These are esters of phosphatidic acid and nitrogen-containing alcohols, namely choline in lecithins and either ethanolamine or serine in the cephalins.

Lecithins — phosphatidyl cholines

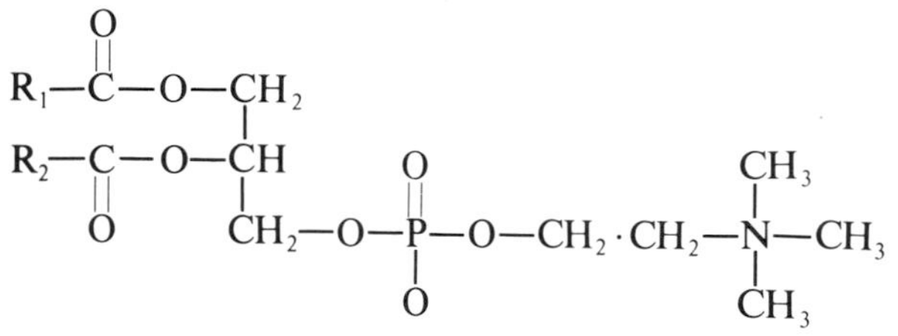

$$R_1-\overset{\overset{O}{\|}}{C}-O-CH_2$$

$$R_2-\overset{\overset{}{|}}{\underset{\overset{}{O}}{C}}-O-\overset{}{CH}$$

$$CH_2-O-\overset{\overset{O}{\|}}{\underset{\overset{}{O}}{P}}-O-CH_2\cdot CH_2-\overset{\overset{CH_3}{|}}{\underset{\overset{}{CH_3}}{N}}-CH_3$$

This is the most abundant polar lipid in both plants and animals.

Cephalins — phosphatidyl ethanolamines and serines.

Phosphatidyl ethanolamine

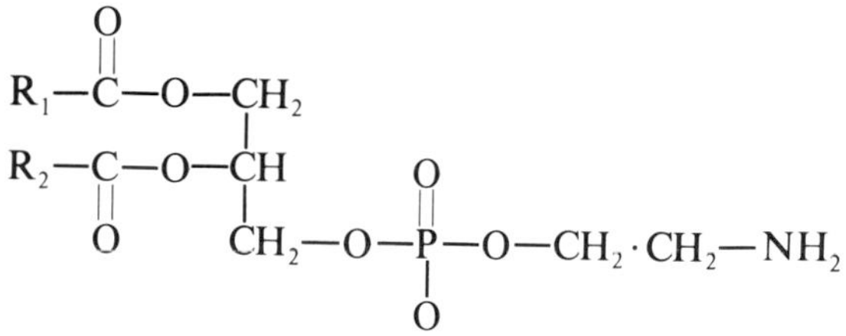

$$R_1-\overset{\overset{O}{\|}}{C}-O-CH_2$$

$$R_2-\overset{\overset{}{|}}{\underset{\overset{}{O}}{C}}-O-\overset{}{CH}$$

$$CH_2-O-\overset{\overset{O}{\|}}{\underset{\overset{}{O}}{P}}-O-CH_2\cdot CH_2-NH_2$$

Phosphatidyl serine

$$R_1-\overset{\overset{O}{\|}}{C}-O-CH_2$$

$$R_2-\overset{\overset{}{|}}{\underset{\overset{}{O}}{C}}-O-\overset{}{CH}$$

$$CH_2-O-\overset{\overset{O}{\|}}{\underset{\overset{}{O}}{P}}-O-CH_2CH_2-\overset{\overset{NH_2}{|}}{C}OO$$

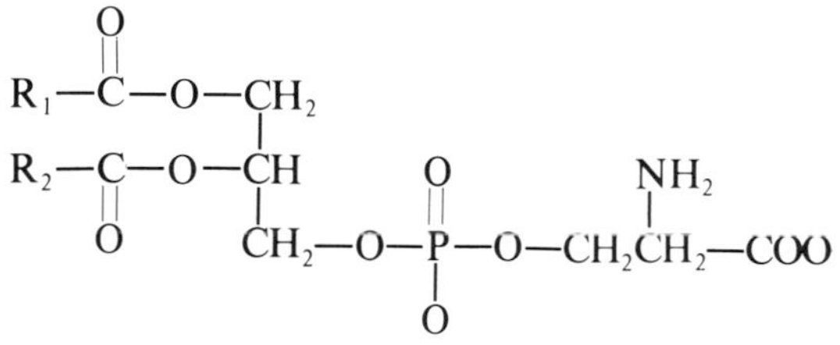

These are often the major lipids in bacteria. Their constituent fatty acids tend to be more saturated than those of the glycerol and cholesterol esters.

Plasmalogens.

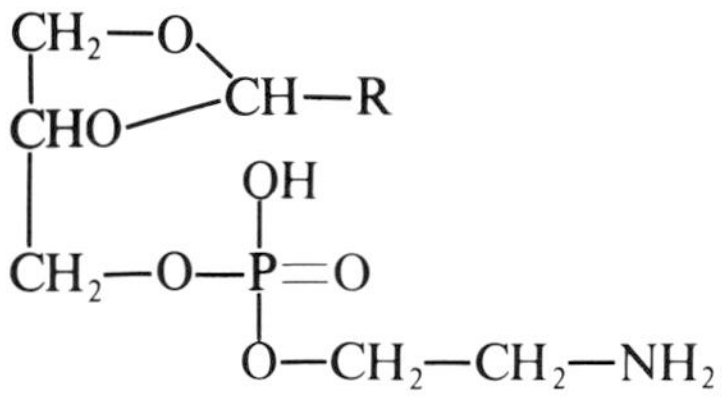

$$CH_2-O$$
$$CHO \quad CH-R$$
$$CH_2-O-\overset{\overset{OH}{|}}{P}=O$$
$$O-CH_2-CH_2-NH_2$$

These phosphatides contain an acetal linkage and have a fatty aldehyde radicle in place of the customary fatty acid. They are present in small but significant amounts in myelin, heart muscle, and especially in the tissues of ruminant animals.

Cardiolipins – diphosphatidyl glycerols

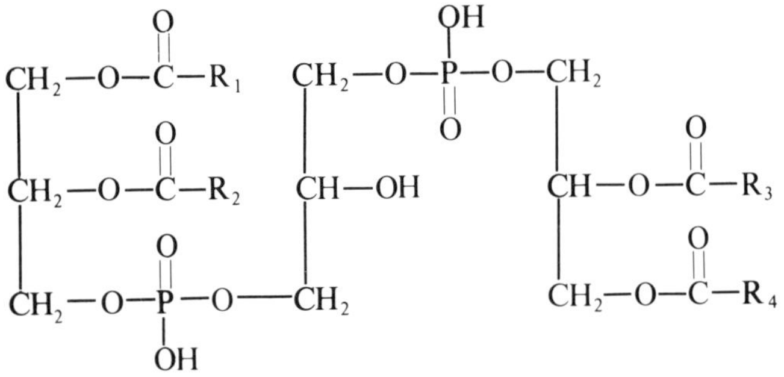

$$
\begin{array}{lll}
CH_2-O-\overset{\overset{\displaystyle O}{\|}}{C}-R_1 & CH_2-O-\overset{\overset{\displaystyle OH}{|}}{\underset{\underset{\displaystyle O}{\|}}{P}}-O-CH_2 & \\[2em]
CH_2-O-\overset{\overset{\displaystyle O}{\|}}{C}-R_2 & CH-OH & CH-O-\overset{\overset{\displaystyle O}{\|}}{C}-R_3 \\[2em]
CH_2-O-\overset{\underset{\displaystyle OH}{|}}{\underset{\overset{\displaystyle O}{\|}}{P}}-O-CH_2 & & CH_2-O-\overset{\overset{\displaystyle O}{\|}}{C}-R_4
\end{array}
$$

Cardiolipins occur throughout nature and include mainly unsaturated fatty acids. These are the main lipid component of mitochondrial membranes and although there is no specific histochemical method as yet available for their identification, they may well be demonstrated incidentally by certain methods. One of the few antigenic lipids, this is the factor in beef heart that is responsible for the serological test for syphilis.

2.4. Sphingolipids

These lipids contain a long chain base–sphingosine. The simplest is ceramide, a fatty-acid amide of sphingosine, the basic structural unit upon which all spingolipids are built.

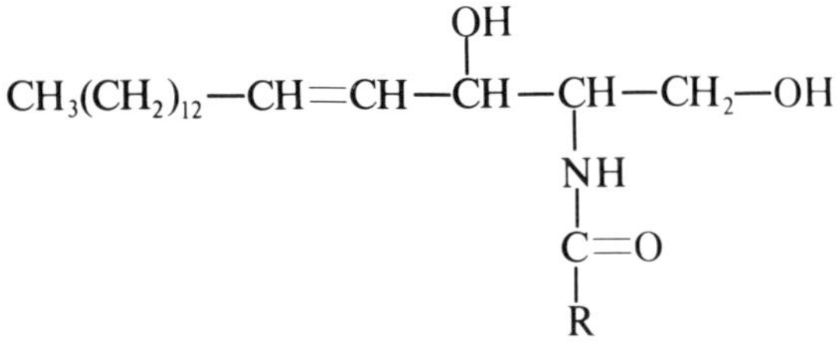

$$
CH_3(CH_2)_{12}-CH{=}CH-\overset{\overset{\displaystyle OH}{|}}{CH}-\underset{\underset{\underset{\underset{\underset{\displaystyle R}{|}}{C{=}O}}{|}}{NH}}{CH}-CH_2-OH
$$

Ceramide

Sphingomyelins

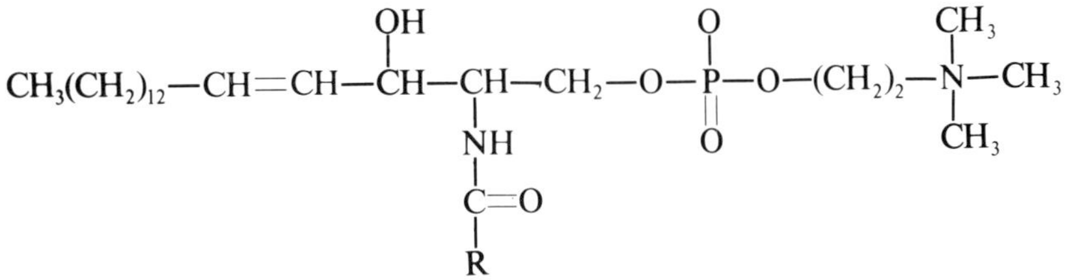

$$
CH_3(CH_2)_{12}-CH{=}CH-\overset{\overset{\displaystyle OH}{|}}{CH}-\underset{\underset{\underset{\underset{\underset{\displaystyle R}{|}}{C{=}O}}{|}}{NH}}{CH}-CH_2-O-\overset{\overset{\displaystyle O}{\|}}{\underset{\underset{\displaystyle O}{\|}}{P}}-O-(CH_2)_2-\overset{\overset{\displaystyle CH_3}{|}}{\underset{\underset{\displaystyle CH_3}{|}}{N}}-CH_3
$$

Sphingomyelins are esterified with phosphoryl choline at the C_1 position on the ceramide group and have also a single fatty acid which is usually either stearic or lignoceric acid, both of which are fully saturated and linked by an amide bond at the C_2 position. These are basic lipids.

Cerebrosides

These glycolipids have a hexose molecule combined as a glycoside with an OH group of the ceramide skeleton. Glucose is the typical sugar in both plant and animal cerebrosides, with the exception of mammalian brain where the major cerebroside is ceramide galactoside.

Sulphatides

These lipids are sulphate esters of cerebroside and can only be detected microscopically when present in abnormal amounts in sulphatide storage disease. These are the only lipids which are sufficiently acidic to induce a metachromatic shift in the basic aniline dyes such as cresyl violet.

Gangliosides

Gangliosides are so-called because they occur principally in ganglion cells of the nervous system. They are the most complex of all the glycolipids and differ from cerebrosides in that they contain additionally n-acetyl neuraminic acid (sialic acid).

n-acetyl neuraminic acid

Several types of ganglioside have been characterized in terms of their hexose moieties and the number of sialic-acid residues they contain; e.g. monosialo ganglioside (GM_1) which accumulates abnormally in neurones in cases of infantile amaurotic idiocy. The gangliosides themselves are water-soluble but those that accumulate in the gangliosidoses are largely bound to protein and consequently resist aqueous and possibly also some organic solvents. These are the only lipids to date that have been demonstrated immunohistochemically (see Chapter 9).

2.5. Lipofuscins

Although not strictly lipids, these yellow-brown pigments must be included here because they comprise peroxidized and subsequently polymerized derivatives of certain unsaturated lipids. In spite of their resistance to organic solvents they retain some staining characteristics of their lipids of origin, although this will vary according to their degree of peroxidation. Ceroid is an early version of lipofuscin and can be classed as a Sudanophilic substance, whereas at later stages in the oxidation process the lipopigment is no longer Sudanophilic but instead becomes autofluorescent, acid-fast, and able to reduce Schmorl's ferric—ferricyanide reagent.

Lipopigments are intralysosomal and are predictably found at sites of lipid deposition following, for example, fatty degeneration of liver and heart muscle. Intraneuronal lipofuscin granules which accumulate during life as the so-called 'wear and tear' pigment, are not necessarily pathological, unlike the ceroid—lipofuscin material that features in certain metabolic storage disorders, with a heterogeneous but distinctive ultrastructure when viewed electron-optically (see Fig. 6).

The physiological roles of lipids

In pathology it is only too easy to regard lipids as abnormal and to overlook their roles as essential constituents of normal tissues (Table 2). Free fatty acids are the basic building blocks from which, together with other radicals, complex lipids are constructed. Although certain fatty acids can be synthesized *in vivo* from carbohydrates, others — the so-called 'essential' fatty acids — must be acquired directly from the diet. These polyunsaturated fatty acids include linoleic and arachidonic acids which are thought to protect against cholesterol deposition in the cardio-vascular system. Cod-liver and sunflower-seed oils, for example, are therefore recommended in place of their more saturated counterparts, found in animal fats and dairy produce. Dietary lipids comprise mainly esters of cholesterol and glycerol which are derived chiefly from meat and milk products. In non-ruminant animals such esters are split by pancreatic enzymes — the esterases and lipases — and are emulsified by bile salts before being absorbed by the intestinal mucosa for re-esterification within mucosal cells for transfer to the liver via the thoracic duct and vena cava. Dietary fats are essential for the absorption of all vitamins except those of the B group which are water-soluble.

To facilitate intravascular transport, cholesterol and glycerol esters are conjugated with a protein vehicle incorporating phospholipids, and they circulate in this soluble lipoprotein state to supply tissue requirements.

The variety of sterols found in nature can be either free or esterified, with marked diversity among the lower forms of life. Ergosterol, for example, is typical of the fungi, fucosterol is typical of brown algae, and sitosterol appears in many plant species. Cholesterol is the principal sterol in animal tissues and may be obtained either from the diet or by endogenous synthesis, chiefly within the liver.

Cholesterol itself is an important precursor of bile acids and adrenal steroid hormones, but above all serves as a major consituent of the plasma membrane where, in equimolar combination with phospholipids, these lipids are intimately associated with protein in a characteristic 'bimolecular leaflet' formation. The myelin sheath is uniquely rich in lipids since it is constructed solely from compacted membrane, produced by the Schwann cell in the peripheral nerve and by the oligodendrocyte in the central nervous system, and wrapped spirally around axons during the process of myelination. The stable lamellar structure of myelin is due to the alternate layering of cirumferentially arranged protein with radially aligned lipids.

Cholesterol esters and waxes such as those secreted in sebum and lanolin (the grease from sheep wool which is commercially useful in the cosmetic industry) provide a protective, water-repellent coating for hairs. This protection is especially

Table 2. *Physiological roles of lipids*

Fatty acids	Building blocks for complex lipids
	Precursors of prostaglandins
Cholesterol	Precursor of steroid hormones and bile salts
Cholesterol with phospholipids	Major components of cell membrane therefore especially concentrated in the myelin sheath
Cholesterol esters and waxes	Water-repellent coating for fur, feathers, insects, and plants
Glycerol esters	Long-term energy reserves, provide thermal insulation and cushion delicate organs
Phospholipids	Assist intravascular transport of cholesterol in various lipoprotein forms
	Encourage alveolar expansion in the fetal lung

valuable to those animals most exposed to rain, such as Lake District sheep, and is essential for aquatic animals. The beaver, for example, possesses two inguinal sacs which secrete castoreum, a greasy substance for grooming his fur, perhaps analogous to the supply of fatty acids in avian preen glands for waterproofing feathers.

Waxes are similarly used to coat insect wings, flower petals, leaves, and micro-organisms, not only to impede entry of water but also to prevent the loss of intra-cellular moisture. Beeswax is structurally important in the fabrication of a residential 'comb' for the honey bee. Waxes may also be utilized for energy reserves in the plant kingdom and in the sperm whale in place of the glycerol esters that customarily provide the long-term energy store in animals, avian yolk sacs, and in certain plant seeds.

Triglycerides have functions other than to supply energy. They provide thermal insulation when laid down as adipose tissue in warm-blooded land animals or as blubber in their marine counterparts. Adipose tissue in mammals additionally serves as a protective cushion around delicate organs such as the kidney, to resist mech-anical injury. Like skeletal muscle and liver oils throughout the fish kingdom, the blubber of whales and dolphins is believed to contribute to their bouyancy.

We have already seen that phospholipids in combination with cholesterol and proteins, constitute the plasma membrane which is so vital to all cells and which in compacted form constitutes the myelin sheath. Mitochondrial membranes are almost exclusively comprised of phospholipids, unlike cell membranes themselves wherein cholesterol and phospholipids are in equimolar combination.

Phosphoglycerides are secreted by Clara cells in bronchiolar epithelium and may be important at birth to facilitate the expansion of alveoli in the fetal lung. The surface-active, detergent properties of phospholipids have already been alluded to on account of their contribution to the lipoprotein molecule, assisting in the intra-vascular transport of cholesterol and triglycerides.

It will be appreciated that the subject of lipids is relevant not only to biologists, but also holds interest for botanists and food and cosmetic scientists, among other microscopists. However, it is probably the field of pathology that lipid histochem-ical techniques are most frequently utilized.

Applications of lipid histochemistry to pathology

Having outlined some of the beneficial effects and essential roles that lipids may play under normal circumstances, let us now consider some of the pathological situations in which lipids are implicated.

Since the myelin sheath is uniquely rich in cholesterol and phospholipids, being composed entirely of cell membrane, it is not therefore surprising that lipid histochemistry finds its major application in diseases of the nervous system. Consequently, both myelin degeneration and the metabolic storage disorders affecting lipids will be considered here in some detail. Other applications that lipid techniques might have outside the nervous system will be listed briefly afterwards.

4.1. Demyelination

Secondary demyelination is an inevitable consequence of brain damage due to infarction, haemorrhage, and tumours of the CNS, or when peripheral nerves are severed (Wallerian degeneration). In these cases both the axon and its myelin sheath degenerate. In *primary* demyelination, on the other hand, when the myelin sheath is selectively destroyed by toxins, infectious agents, or allergic influences, the axon may largely be spared. Discrete foci of demyelination usually result, such as the typical plaques seen in cases of multiple sclerosis. However, whether primary or secondary, in terms of lipid histochemistry the two types of lesion are essentially the same, although it may be important to distinguish between actively demyelinating lesions and old lipid-free scars. Morphological changes in a demyelinating lesion are accompanied by qualitative changes in the myelin lipids themselves as the hydrophilic myelin fragments from the disrupted sheath are phagocytosed by microglia in the CNS, or by Schwann cells in the PNS. Within these cells, free cholesterol is converted to cholesterol esters which, in contrast to normal myelin, are hydrophobic and therefore intensely Sudanophilic. This physico-chemical difference provides the basis of several methods for the simultaneous demonstration of normal and degenerating myelin (see p. 42).

4.2. Lipid storage disorders

Perhaps the most valuable service that lipid histochemistry can provide is in the diagnosis of the metabolic storage disorders involving lipids, which may therefore be worth considering in some detail. These rare inherited diseases are due to the

Table 3. *Lipid storage diseases*

Disease	Enzyme defect	Lipid stored	Biopsy source	Histochemistry	Comments
GM_1 gangliosidosis	Ganglioside-β-galactosidase	GM_1 ganglioside	Rectum Appendix	BHPS Thionin pH3	Acute infantile onset Mental retardation; short survival
GM_2 gangliosidoses Tay Sachs' disease	Hexosaminidase A	GM_2 ganglioside	Rectum	BHPS	Mental retardation and blindness
Sandhoff's disease	Hexosaminidase A + B	GM_2 ganglioside	Appendix	Thionin pH3	Similar but rapidly progressive
Batten's disease	?	Ceroid-lipofuscin	Rectum Appendix	Long Sudan Black; PAS; thionin pH3; long ZN; LFB; autofluorescence; EM	Short survival Therapy: restrict Vitamin A
Farber's lipogranulomatosis	Acid ceramidase	Ceramide	Skin	PAS; EM	Fatal during second year of life
Fabry's disease	Ceramide trihexosidase and α-galactosidase	Ceramide trihexoside	Skin Kidney	PAS; Sudan black B Polarized light; PAN	Purple skin rash; renal failure; burning pains in extremities
Refsun's syndrome	Phytanic acid hydrolase	Phytanic acid	Liver	Copper rubeanic acid*	Therapy: restrict chlorophyll
Wolman's disease	Cholesterol esterase	Cholesterol esters and triglycerides	Liver Lymph node	PAN Calcium-lipase	Fatal during first year
Cholesterol ester storage disease	Cholesterol esterase	Cholesterol esters and triglycerides	Liver Lymph node	PAN Calcium-lipase	More benign
Tangier disease	Impaired synthesis of apoprotein A	Cholesterol esters	Tonsils	PAN* PAN + Digitonin	Follows a fairly benign course

Niemann–Pick's disease	Sphingomyelinase	Sphingomyelin and cholesterol	Bone marrow Liver; rectum	NaOH–DAH NaOH–FeH; PAN	Heptatosplenomegaly and mental retardation
Gaucher's disease	Glucocerebrosidase	Glucocerebroside	Bone marrow Liver	Modified PAS (+ diastase if liver)	Hepatosplenomegaly; mental retardation in infantile form only
Krabbe's leuco-dystrophy	Galactocerebroside-β-galactosidase	Galactocerebroside	Brain	Modified PAS	Multinucleate 'globoid' cells around vessels; mental retardation; fatal by two years
Metachromatic leucodystrophy	Arylsulphatase A	Sulphatide	Urine deposit Sural nerve	Acriflavine–DMAB Toluidine blue–acetone; High iron diamine	Loss of myelin; mental retardation; fatal during first decade

*Predicted not tested.

complete or partial deficiency of certain enzymes — usually lysosomal — associated with lipid catabolism, resulting in an accumulation of the relative lipid metabolites at the defective stage in their metabolic cycle. In the leucodystrophies the cells of the mononuclear phagocyte (MP) series may be involved systemically, whereas lipid accumulation in the neurolipidoses is principally within neurones. The storage diseases have different ages of onset and variable clinical and pathological manifestations but most of them are fatal during the first decade of life.

The most significant group of these lipid storage disorders are the sphingolipidoses. Gangliosides, for example, are the lipids responsible for the intraneuronal accumulations in Tay Sachs' disease and the other GM_1 and GM_2 gangliosidoses (plate, bottom right inside cover). They may be additionally implicated in neuronal changes secondary to systemic disturbance of mucopolysaccharide metabolism in the three types of glycosaminoglycan storage disease.

A further important group of neuronal storage diseases — collectively termed Batten's disease — are the ceroid lipofuscinoses, the most commonly occurring neurodegenerative disorders in childhood.

It is customary nowadays to investigate suspected cases of neurolipidosis with appendix or rectal biopsy (Bodian and Lake 1963) because ganglion cells in the gastrointestinal tract conveniently mimic the storage patterns of their CNS counterparts (see Fig. 6, p. 56). Lipid storage diseases affecting the MP system on the other hand, require different biopsy material as indicated in Table 3. Such diseases include Gaucher's (cerebroside) and Niemann Pick's (sphingomyelin and cholesterol), while sulphatide is the lipid implicated in the condition known as metachromatic leucodystrophy (see Fig. 5, p. 52).

Another metabolic disorder — Wolman's disease — results in the systemic accumulation of cholesterol and glycerol esters within macrophages. Table 3 indicates the enzymes responsible for each of the storage diseases, the lipid stored, the biopsy site, and the histochemical method of choice to localize and identify the lipid in question.

It should be pointed out that more direct and precise methods are now available whereby the underlying enzyme deficits in these conditions may be detected. Enzymic assays of leucocytes, cultured fibroblasts, amniotic fluid cells, and even babies' tear drops (!) have proved successful. Without a known cure for these conditions, such techniques are particularly valuable to screen carriers of a defective gene and to test 'at risk' pregnancies by means of amniocentesis so that an affected fetus may be aborted. Subsequent genetic counselling in these cases aims to reduce further expression of an inherited defect.

For maximum effect, one would approach these diagnostic problems from several angles, combining electron microscopy, thin-layer chromatography of tissue extracts, enzyme assay, and lipid histochemistry. Where facilities for biochemical investigations are beyond the scope of the microscopy laboratory, histochemistry may well provide the sole means of diagnosis. Indeed it has a distinct advantage over the other disciplines — namely the ability to localize metabolites within individual cells and even, occasionally, to subcellular organelles with the electron

Table 4. *Further applications of lipid histochemistry in general pathology*

Tissue	Condition	Lipid involved	Reference
Brain	Parkinson's disease	Sphingomyelin in Lewy bodies in substantia nigra	den Hartog Jager (1969)
	Pick's presenile dementia	Ganglioside in neurones	de Groot and den Hartog Jager (1980)
Eye	'Arcus senilis'	Cholesterol and glycerol ester accumulation in the cornea	
Mouth	Dental cysts	Cholesterol and its esters evoking an encapsulated granulomatous response	
Lung	Lipoid pneumonia due to accidental inhalation of oily foodstuffs, e.g. milk and cod-liver oil-, oil-based drugs or deliberate introduction of oil-based contrast media.	Granulomatous and fibrous tissue containing persistent droplets of Sudanophilic lipid	
	Paraffinoma due to aspirated liquid paraffin	Liquid paraffin in extracellular vacuoles is a saturated hydro-carbon oil which stains with oil red O but not with Sudan black B, OsO_4 and other lipid methods.	
	Fat embolus lodged characteristically in the pulmonary tree	Triglycerides if embolic material is derived from bone marrow fat; alternatively cholesterol esters if atheromatous in origin	
Skeletal muscle	Lipid myopathy	Neutral fat droplets in Type I fibres may be stained with oil red O and Sudan black B	
Heart	Cardiomyopathies induced by alcohol, other toxins, and viruses	Triglyceride deposition in cardiac muscle	
	Thrush-breast heart in severe anaemia	Focal deposition of triglycerides as heart muscle is deprived of O_2	
	Brown atrophy	Lipofuscin accumulation in the senile heart	
Arteries	Atherosclerosis	Cholesterol and esters deposited intramurally in plaques can be demonstrated with the fat stains and viewed in polarized light to distinguish the crystalline element. Alternatively, the extent of atheroma can be seen in the gross when whole segments of artery are stained with oil red O	

Table 4. (*Continued*)

Tissue	Condition	Lipid involved	Reference
Liver	Fatty change. Although a feature of all O_2-deprived cells, the liver is especially prone to fatty degeneration following hypoxia, alcoholism, protein deficiency, etc.	Triglyceride accumulation in parenchymal cells can be demonstrated with any of the fat stains	
Tendons and fascia	Ageing	Free and esterified cholesterol accumulates in these avascular sites	Adams *et al.* (1974)
Skin	Xanthoma	Free and ester cholesterol in subcutaneous granulomata, especially over joints	
Kidney	Nephrotic syndome	Triglycerides in diffuse stippling in renal tubules; free fatty acids in dense focal masses, some extracellular in parenchyma (see top left plate inside back cover)	Archibald and Orton (1970)
	Bright's disease (membrano-proliferative glomerulonephritis).	Sudanophilic lipid localized in tubular epithelium in the original specimen of this condition (1827).	
Miscellaneous			
	Whipple's disease. This rare infection is due to an organism with an incomplete coat, hence difficult to identify	Triglycerides and free fatty acids accumulate in the lymphatics	
	Reye's disease — encephalopathy and fatty degeneration of the viscera	Triglycerides appear systemically	Reye *et al.* (1963)
	Cholesterol resorption observed in macrophages *in vivo* by implanting crystals subcutaneously, and *in vitro* by applying cholesterol to macrophages in culture	Cholesterol can be distinguished from its Sudanophilic esters produced in macrophages, if a section stained with oil red O is viewed in polarized light	Bayliss High (1976).
	Tuberculosis	Lipid capsule of *mycobacterium tuberculosis* is stained by carbol fuchsin in the Ziehl—Neelsen method	

microscope (see Chapter 8). However, in view of the limitations and artefacts experienced with lipid methods (see §5.1), it is imperative that histochemical observations are considered in conjunction with clinical and morphological evidence, hopefully so that a characteristic pattern can be attributed to a particular pathological condition.

Undoubtedly, microscopists in fields other than pathology will encounter different problems that require to be solved histochemically. Table 4 indicates examples of the application of lipid histochemistry to specific biological and pathological problems. The topic is treated superficially; the list is merely illustrative and is derived chiefly from work in this department.

5

Preparation of tissues for lipid histochemistry

5.1. Fixation

Before a histochemical preparation can be properly evaluated, it is essential to know that lipids have actually been retained in the section. The choice of fixative is governed by a variety of factors. Unfixed sections theoretically provide the ideal substrate for histochemical reactions, giving the most authentic picture of the lipids in their *in vivo* condition. In practice, however, some lipids tend to be soluble in the staining reagents so that some degree of preservation is desirable if lipids and the sections themselves are to withstand potentially destructive staining procedures.

Ideally, a fixative should preserve tissues against bacterial putrefaction and the autolytic action of lysosomal enzymes released post-mortem. The fixative must be buffered to reduce osmotic distortion and retain the shape and volume of the cells during subsequent processing and staining. Most important of all histochemically, fixation should prevent the loss of compounds to be demonstrated microscopically without altering their reactive groups. However, there are only two reagents that are capable of fixing lipids in the true sense of the word, so that they become insoluble in organic solvents. These reagents are OsO_4 and chromic acid, both of which substantially alter the chemical reactivity of the lipids they fix.

Lipids are not actually 'fixed' by formaldehyde as proteins are, but they can be better retained in the section when held in a surrounding matrix of coagulated structural and soluble tissue proteins. Formalin has little effect on neutral fats, though some may crystallize during prolonged storage and thereby lose their original physical characteristics. However, formaldehyde can become oxidized to formic acid during storage; thus it is imperative to buffer fixative solutions to avoid an acid medium that would be conducive to lipid hydrolysis. Lillie regarded 2 per cent calcium acetate to be as effective as phosphate buffer in maintaining the neutrality of formalin, with the advantage that calcium ions protect phospholipids (see below and Fig. 1) and therefore calcium–formal is widely used as a general fixative for lipid histochemistry.

Formaldehyde itself can adversely affect some phospholipids — notably plasmalogens and lecithins which are readily degraded to their water-soluble derivatives. Plasmologens fortunately can be demonstrated in unfixed sections because the initial stage of their staining technique is mercuric chloride hydrolysis and the mercury salt acts as a fixative in its own right. Likewise one could use unfixed

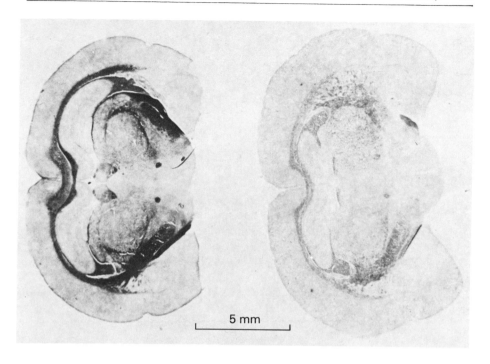

Fig. 1. Normal rat brain sections stained with the standard Sudan black B method after six months in calcium–formal (left) and formal–saline (right).

material for the dichromate–acid haematein method for choline-containing phospholipids (see later), since the first step in this procedure entails chromation in the presence of calcium ions. On the other hand, if phosphoglycerides arc to be demonstrated by the gold hydroxamic acid method, tissues must be adequately fixed beforehand so that sections may withstand the alkaline reagents employed thercin. Tissues destined for the cholesterol and triglyceride techniques described below, need to be well fixed; happily neither of these methods is impeded by formalin fixation.

Protection of phospholipids during fixation can be achieved by the addition of calcium and cadmium ions to formalin(Baker 1945) without actually altering their solubility in organic solvents. These ions are presumed to be effective due to the formation of a lattice of complex coacervates of phospholipids with proteins or mucins, which remains stable during fixation. However, although the diffusion of lipids into the fixative may be prevented, they remain vulnerable to the subsequent effect of alcoholic dye solvents. Calcium–formal (2 per cent calcium acetate added to 10 per cent formalin) is the preferred fixative for lipid histochemistry. In fact sodium ions are known actually to encourage the solution of brain lipids (as indicated in Fig. 1) so regular formal–saline should be avoided, particularly for neural tissues.

An incidental effect of calcium is to convert free fatty acids to their calcium soaps. This is unimportant if fatty acids are to be demonstrated histochemically but may present a problem should fatty acids require to be extracted selectively (see below) because saponification alters their solubility properties in organic solvents.

Evidently there can be no single mode of fixation to suit all lipids and every technique. Our aim is to achieve an acceptable compromise allowing maximum preservation of tissue architecture with minimal alteration in the lipids themselves.

What this means in practice is that when one is looking for fats it is better to cut sections of fixed material so that fatty droplets such as one finds in degenerating myelin, will be held in a matrix of coagulated protein and should therefore be less likely to float out of the section during staining. Extracellular fat on the other hand, seems to be better retained if unfixed cryostat sections are mounted on to slides so that the fat sticks to the glass.

For the different phospholipid techniques to be described later, cryostat sections are either treated unfixed or post-fixed in calcium–formal, according to individual specifications. With methods that employ particularly destructive reagents, fixation should be extended – say overnight – rather than the customary one hour indicated for 'mild' methods. These directives are listed in Table 5.

5.2. Microtomy

Frozen sections are necessary for lipid histochemistry since routine processing for paraffin and plastic embedding will remove all but a few protein-bound lipids. It is conceded that freezing and thawing is by no means the ideal way to treat tissues destined for microscopy but in this context we are concerned more with the retention of lipids and their identification than with the detailed morphology of the tissue section itself.

Freezing

Specimens of fresh tissue should be speedily frozen for the cryostat either with compressed CO_2, liquid gases such as nitrogen, or perhaps more conveniently with an aerosol spray of dichlorodifluoromethane which is supplied commercially, e.g. '*Cryospray*' (Bright Instrument Company Ltd). The specimen is placed on a cork disc and surrounded with a blob of OCT (Ames Tissue Tek OCT; Miles Labs.), immersed in liquid nitrogen or sprayed with the freezing aerosol, then attached to a metal chuck by an intervening drop of water or OCT, for sectioning in the cryostat. Alternatively fixed material can be frozen directly on to the stage of a thermoelectric freezing microtome, e.g. *Metron* (Frigistor) Ltd. for sectioning in the open laboratory.

Table 5. *Fixation protocol for lipid histochemistry*

Method	Lipid	Fixation
u.v. Schiff	Unsaturated lipids	Unfixed cryostat sections
Plasmal reaction	Plasmalogens	
Bromine–Sudan black B	All lipids	
Sudan black B	Fats and some phospholipids	Frozen sections of short-fixed tissue, or cryostat sections post-fixed for 1 hour in calcium–formal
Oil red O	Fats	
Nile Blue SO$_4$	Fatty acids, phospholipids, and neutral fats	
Acetone–Nile blue SO$_4$	Phospholipids	
Copper–rubeanic acid	Free fatty acids	
Dichromate–acid haematein	Lecithins and sphingomyelins	
Acriflavine–DMAB		
Toluidine blue–acetone	Sulphatides	
High iron diamine		
Borohydride–periodate–Schiff	Gangliosides	
Thionin pH 3		
Perchloric acid–naphthoquinone (PAN)	Cholesterol and its esters	Tissues pre- or post-fixed for at least 18 hours in calcium–formal
Digitonin–PAN	Free cholesterol	
Calcium–lipase	Triglycerides	
Gold hydroxamic acid	Phosphoglycerides	
NaOH–dichromate–acid haematein	Sphingomyelins	
Modified PAS	Cerebrosides	

Supportive media

It is not always necessary to embed tissues in a medium when the intrinsic water acts as its own embedding medium when frozen. However, small and fragmentary tissues require to be protected by an inert, water-soluble surrounding matrix, such as OCT(Miles Labs) before freezing.

Alternatively, gelatine can be used to facilitate the handling of difficult tissue fragments and to assist in their accurate orientation.

Gelatine impregnation

Solution

Gelatine	16 g
Glycerine	15 cm^3
Distilled water	70 cm^3
Thymol	a small crystal

Store at 4°C and melt in a water bath at 37°C for use.

Method

1. Fix tissues in calcium–formal.
2. Wash thoroughly in running water.

3. Impregnate with gelatine—glycerine mixture at 37°C for six hours.

4. Transfer to fresh mixture and embed tissue in moulds which are allowed to set in the refrigerator.

5. Trim blocks and harden them by fixing in calcium—formal overnight before sectioning.

Microtomes

Fixed tissues can be cut on a conventional open freezing microtome with the stage traditionally cooled by a blast of compressed CO_2, but with this method it is difficult to maintain the block at the ideal temperature for sectioning, making it difficult to obtain sections of a consistent thickness. This problem was largely overcome when a thermoelectric module was incorporated into the microtome stage so that the cutting temperature could be controlled electrically and maintained indefinitely. This is achieved by passing a direct current through a series of bimetallic couples to produce the 'Peltier' effect; i.e. one surface cools down at the expense of its opposite surface which generates heat and this heat is removed by a water cooling device. Section cutting on any open microtome is hampered by condensation and is not at all suitable for unfixed material. Sections are floated on to a water bath before being mounted on to slides; alternatively, they can be processed free-floating by transferring them through the reagents by means of glass 'hockey sticks'. Unfixed material would disintegrate under these circumstance and instead should be sectioned in the crysotatic microtome.

The cryostat has revolutionized histochemistry. By housing a conventional rotary or rocking microtome entirely within a refrigerated cabinet, at an adjustable ambient temperature and with all its controls adapted to be operated externally, it is possible to cut and mount sections on to slides in their rigid, frozen state. Sections are kept flat by means of an 'anti-rol' guide-plate. Popular models of the cryostat include those produced by Bright Instrument Co Ltd, SLEE Medical, and Ames Tissue Tek (Miles Labs) and can be either front- or top-opening models.

Although the cryostat was originally devised to provide unfixed sections for enzyme and immunocytochemical techniques, it has proved equally invaluable for other disciplines including lipid histochemistry and can also be used to cut fixed tissue with greater speed and consistency that with the traditional freezing microtome. A further advantage of cryostat sections for lipid histochemistry is that fat globules seem less likely to be washed out of sections if they have been mounted on to slides in the fresh state.

Sectioning

The optimum temperature for sectioning fresh frozen material varies for different tissues. As a generalization, soft tissues such as brain and kidney cut best at around − 15°C whereas tougher material such as skin and breast needs to be at about − 25°C, as do specimens that contain much frank fat. Fixed material should be sectioned at temperatures higher than those specified for fresh tissues, say around $-10°C$. In general, soft tissues should be cut slowly but hard tissue seems to cut better at a faster rate. Fresh tissues are generally prepared at $15 \mu m$ for lipid methods, to be roughly comparable with 10μ sections of fixed material.

Adhesives

When formalin-fixed tissues are sectioned in the cryostat, sections are less adhesive to slides than those prepared from fresh tissue because their sticky proteins have been denatured during fixation. Chrome-gelatin-coated slides are therefore recommended for fixed sections to prevent their detachment from slides, especially during passage through alkaline reagents. Simply dip clean microscope slides in a solution of 1 per cent gelatin containing a pinch of chromium potassium sulphate and allow the slides to drain dry before applying sections.

6

Control sections

In view of the considerable limitations encountered in lipid histochemistry, the importance of control sections cannot be overstated. Reactions that are specific for a particular chemical group may be selective for a single member of the lipid family but could equally include the same reactive group in non-lipid tissue constituents. For example, the 1,2-glycol grouping responsible for the PAS (Periodic acid–Schiff) positivity of cerebroside, also depicts the hexoses in glycosaminoglycans generally. Not only do we need to exclude interference from cross-reacting non-lipid compounds, but we should also verify the authenticity of a reaction by comparison with a 'positive' control section processed in parallel.

6.1. Positive controls

Tissues that are known to contain certain lipids can be prepared as a composite block and multiple cryostat sections cut and stored at $-20°C$ until required. Myelin is of course a good source of all the phospholipids together, but it will be appreciated that control material for individual phospholipids can be provided only rarely from proven cases of the lipid storage disorders. So whenever such material becomes available at biopsy or post-mortem, it should be stored expressly for this purpose. For other lipids, control tissues are suggested in Table 6. Alternatively, some lipids are available commercially (e.g. from Sigma Chemical Co, Ltd) and can be dissolved in chloroform, applied to filter paper, and processed as a tissue section. Indeed this technique has been used to validate all the methods described in this handbook (see Fig. 2).

6.2. Negative controls

Delipidized sections play a crucial role in lipid histochemistry and should be included routinely along with normal sections for all methods. Lipids are deliberately extracted in order to verify that the reaction product under consideration in the normal section can be attributed to lipid, or, on the other hand, to assess the extent of any interference from non-lipid groups that may react identically.

Delipidization. A total lipid extraction can be achieved with chloroform: methanol (2:1) if 1 per cent HCl is included in the solvent to disrupt lipoprotein bonds so that protein-bound lipids may also be extracted. The further addition of a small amount of water (say 4 per cent) is thought to facilitate the removal of phospholipids.

Table 6. *Control tissues*

Control tissues	To demonstrate
Atheromatous artery Adrenal	Cholesterol and its esters
Fatty liver	Triglycerides and free fatty acids
Brain	Most phospholipids
Rat sciatic nerve severed and allowed to degenerate for 12 days *in vivo*	Normal and degenerating myelin lipids

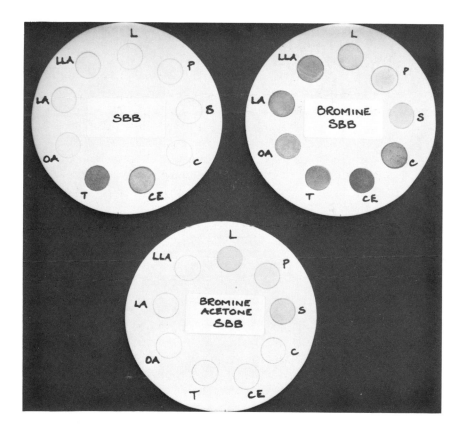

Fig. 2. Spots of pure lipids applied to filter paper and stained as follows. Top left: Sudan black B; top right: bromine – Sudan black B. Key to lipids: L, lecithin; P, plasmalogen; S, sphingomyelin; C, cholesterol; CE, cholesterol oleate; T, triolein; OA, oleic acid; LA, linoleic acid; LLA, linolenic acid.

General lipid solvent

Chloroform	$66 \, cm^3$
Methanol	$33 \, cm^3$
H_2O	$4 \, cm^3$
HCl	$1 \, cm^3$

Use at room temperature for one hour.

Although at one time it was considered feasible to identify individual lipids by their differential solubilities in an elaborate sequence of organic solvents, such procedures have proved to be unreliable — with the sole exception of acetone.

Acetone extraction can be useful to differentiate between phospholipids, which are acetone-insoluble, and fats and cholesterol which are readily soluble in this solvent. It will be seen that this property can be exploited in certain histochemical techniques, when interfering lipids are extracted with acetone so that only the acetone-insoluble phospholipids will be retained in the section for subsequent staining (Fig. 2). Acetone is used at 4°C for 20 minutes and must be anhydrous otherwise a significant amount of phosphoglycerides will be extracted along with the intended apolar lipids.

Limitations and artefacts

Of all histochemical disciplines, that of lipids is the one most prone to artefacts and it is important to understand such limitations in order to interpret histochemical reactions confidently. The following are selected examples of such artefacts, to illustrate the sort of pitfalls that tend to await the lipid histochemist.

7.1. Physical limitations

The first problem is that tissue lipids do not necessarily obey the rules of pure lipids that have been isolated in the test-tube. Their solubility characteristics may be modified by even traces of other lipids. Furthermore, much of the lipid in tissues is present as lipoprotein or proteolipid, bound to, and possibly masked by, proteins and carbohydrates.

We have already seen that formalin fixation can alter physical properties and thereby the solubility of certain fats, notably free fatty acid which may become irreversibly polymerized. Triglycerides may crystallize and lose their Sudanophilia. Although prolonged fixation results in the loss of certain myelin lipids, others in contrast become more resistant to extraction with organic solvents.

The melting point of a lipid will determine its behaviour towards organotropic dyes and it should be remembered that although certain cholesterol esters may be liquid *in vivo*, they can be crystalline at staining temperature and therefore no longer Sudanophilic. In addition, free fatty acids and lecithins are readily soluble in 70 per cent ethanol and may be extracted by dyes that are applied in this solvent (see Fig. 2).

7.2. Chemical considerations

A common problem in lipid histochemistry is interference from cross-reacting non-lipid material, which must be excluded by comparison with a delipidized control section. A good example of such interference is the metachromatic reaction of mast-cell granules which mimics the metachromasia displayed by intralysosomal sulphatide accumulations in metachromatic leucodystrophy.

Perhaps the most notorious artefacts are those due to calcium because calcium deposits are commonly found in association with lipid accumulation, as in atherosclerotic lesions. In techniques that employ heavy metal salts such as copper and lead to demonstrate lipids, these metals will base-exchange with calcium ions to produce a reaction that may be mistakenly attributed to lipid unless a delipidized control section has been included for reference.

Another artefact induced by calcium is its effect on free fatty acids. During fixation in calcium—formal, fatty acids are converted to their calcium soaps and although this should present no problem in methods for depicting fatty acids themselves, it would pose a problem when fatty acids require to be selectively extracted with acetone from sections destined to be stained for phospholipids (see later). Calcium soaps are insoluble in acetone and must be desaponified with hydrochloric acid before they can be successfully extracted, otherwise their reaction with Nile blue SO_4 for example, could be erroneously assumed to be due to phospholipids.

Such potential hazards must be recognized if histochemical reactions are to be interpreted correctly. In view of the limitations to lipid techniques, the approach to a difficult diagnostic problem should ideally be multidisciplinary so that histochemistry can be backed up by thin-layer chromatography of the tissue extract, together with electron microscopy where indicated, in order to reinforce the impression gained microscopically about the identity of a particular lipid.

Microscopic techniques for tissue lipids

Histologists are mainly concerned with morphological change and generally rely on haematoxylin- and eosin-stained sections to relate disturbances in tissue architecture to pathological processes. Sometimes one can discern clefts and vacuoles in paraffin and resin sections that may be assumed to have contained crystalline or fatty lipids prior to their extraction during processing. Nevertheless, a histochemical approach would be necessary to verify such an assumption and to determine the identity of the lipids involved.

8.1. Histophysical methods

Important differences in the physical characteristics of lipids have already been stressed, notably crystalline versus the liquid state and hydrophobia versus hydrophilia. Such differences enable us to discriminate microscopically between certain lipid classes in two important ways.

Birefringence of lipids

Lipids cannot be classified solely by means of their optical properties; nevertheless, it can sometimes be helpful to view a section in polarized light after it has been stained with a dye such as oil red O. The unstained crystalline lipid will appear birefringent (anisotropic) whereas the red-stained fats are isotropic (non-birefringent) (Fig. 3). This property cannot be relied upon to identify individual lipids since it is their melting points within the classes that determine their optical behaviour, but in practice a birefringent lipid is likely to be either free cholesterol or its more saturated esters. However, other lipids may retain some crystalline structure and exhibit conic focal anisotropism — 'Maltese Cross' birefringence. The formation of liquid crystals is dependent on temperature and the presence of water, and they usually comprise cholesterol mixed with phospholipids. It should be remembered that non-lipid crystalline structures such as amyloid are also birefringent.

Fat stains

A series of Sudan dyes are the most popular reagents for detecting lipids in tissue sections, although in fact they demonstrate only those lipids that exist as fats at staining temperature, namely the oily, greasy, hydrophobic lipids; those that are in

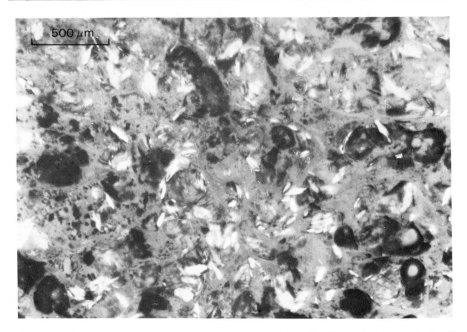

Fig. 3. Fatty liver from a cholesterol-fed rabbit stained with oil red O and photographed with partly crossed polars so that unstained cholesterol crystals appear birefringent and can be distinguished from stained fatty cholesterol esters.

the solid or crystalline state will remain unstained.The surface tension of any fat at a lipid—water interface makes it assume a globular shape, impervious to aqueous reagents but with an affinity for the organotropic Sudan dyes. These dyes are effective because they are more soluble in tissue fats than in their own solvents. Sudan III was the first fat stain to be used for microscopy in 1896 but this and other Sudan dyes such as Sudans I, II, brown, blue, and green are all inferior to Sudan IV. However, oil red O is considered to be even more precise and sensitive than any of these dyes.

In order to penetrate fats, the Sudans must be dissolved in organic solvents, although the latter must be sufficiently dilue to avoid extracting the lipids themselves. The preferred solvent for oil red O is 60 per cent isopropanol whereas 70 per cent ethanol is favoured for Sudan black B. Nevertheless, one can expect only an unsatisfactory compromise between maximum staining intensity and minimal lipid loss. There remains a tendency for free fatty acids and phosphoglycerides to be extracted during staining in alcoholic solutions; furthermore most Sudan dyes fail to stain phospholipids and free cholesterol. In spite of these considerable limitations, the Sudans should not be dismissed as obsolete merely because there are better methods now available for identifying individual lipids. Indeed they remain among the most useful stains in histochemistry. Sudan black B in particular, with judicious modifications, can become one of the most versatile and reliable methods for lipids.

Unlike the other fat stains, Sudan black B comprises two distinct fractions; one colours fats blue-black whilst the second component — a basic dye — stains phospholipids grey. The grey reaction can be enhanced as a bronze dichroism if the stained section is viewed in polarized light. In spite of these useful properties, the role of Sudan black is limited by its inability to stain cholesterol and by extraction of lecithin and free fatty acids in the ethanolic dye bath, unless bromine pretreatment is included in the staining technique (Fig. 2). This step ensures that fatty acids and lecithins become resistant to ethanol extraction and, in addition, bromination causes crystalline cholesterol to be converted to its oily bromo-derivatives which are intensely Sudanophilic (Bayliss and Adams 1972). *Bromine – Sudan black B* therefore provides a simple, sensitive method for detecting each of the main lipid classes and is therefore recommended as a quick, preliminary screening test for tissue lipids prior to their identification by other, more specific, histochemical procedures. Furthermore, with an additional stage of acetone extraction, interposed between bromination and Sudan black staining, only acetone-resistant phospholipids will be demonstrated (Bayliss High 1981). Both these procedures are essentially qualitative and cannot be recommended for detailed morphology. To localize fats more precisely, it is better to use oil red O followed by a haemalum nuclear stain and, if the section is viewed in partly polarized light after staining, one can also appreciate the distribution of unstained, birefringent cholesterol (Fig. 3).

Nile blue sulphate is akin to Sudan black B in that it comprises two separate fractions. This provided the first method (Smith 1908) whereby two types of lipid could be distinguished simultaneously with the microscope. The two components of this dye are a red oxazone which dissolves in fats like any Sudan-type dye, and a blue oxazine which is a basic dye reacting with the carboxyl groups of free fatty acids and phospholipids. Nile blue sulphate is therefore a useful preliminary indicator of the type of lipid present in the tissue section. In practice, lipid mixtures tend to be stained an intermediate purple colour so that perhaps the most useful application of this dye is in Dunnigan's (1968) modification. Fatty acids are first extracted with acetone so that Nile blue sulphate-staining will subsequently be confined to phospholipids. As stated in Chapter 5, one must be aware that acetone will fail to extract free fatty acids in this procedure if calcium–formal has been employed as fixative, since calcium soaps of fatty acids are insoluble and would stain as phospholipids unless they have been desaponified beforehand. A fat stain favoured by botanists is *phosphine 3R* which is taken up by neutral fats to give a silvery-white fluorescence in ultraviolet light. An advantage of this dye over the other fat stains is that its aqueous solvent enables small fat droplets to be better retained during staining.

Methods

Method 1. Bromine – Sudan black B method for lipids (Bayliss and Adams 1972)

Fixation and sections. Cryostat sections post-fixed for one hour in calcium–formal; short fixed frozen sections.

Method

1. Mount sections on to slides and allow to dry;
2. Immerse sections in 2.5 per cent aqueous bromine for 30 minutes at room temperature, inside a fume cupboard;
3. Wash in water and treat with 0.5 per cent sodium metabisulphite for one minute to remove excess bromine;
4. Wash thoroughly in distilled water and treat as directed for the regular Method 2. Sudan black B method (see below).

Method 2. Sudan black B method for fats and phospholipids

1. Rinse sections in 70 per cent ethanol;
2. Stain in saturated Sudan black B in 70 per cent ethanol, filtered just before use, for 15 minutes;
3. Differentiate in 70 per cent ethanol for a minute or two, or until a delipidized control section appears essentially colourless;
4. Counterstain nuclei with 1 per cent methyl green (chloroform washed) for five minutes;
5. Wash well and mount in glycerol–gelatin.

Method 3. Bromine–acetone–Sudan black B method for phospholipids (Bayliss High 1981)

1. Treat sections with bromine and metabisulphite as specified above;
2. Wash sections well and allow them to dry;
3. Extract fats with anhydrous acetone for 20 minutes at $4°C$;
4. Proceed with the regular Sudan black B method outlined above;

Results. Sudan black B alone will stain unsaturated cholesterol and glycerol esters blue-black; some phospholipids appear grey and those in myelin additionally exhibit a bronze dichroism in polarized light (Diezel 1957).

Bromination enhances the reaction of these lipids and, in addition, lecithins, free fatty acids, and free cholesterol will be stained with Sudan black B.

After bromination and acetone extraction, Sudan black staining will be confined to phospholipids.

Method 4. Oil red O method for fats (after Lillie and Ashburn 1943)

Fixation and sections. Cryostat sections post-fixed in calcium–formal; short fixed frozen sections.

Preparation of reagent. The working solution is prepared an hour in advance by mixing three parts of a stock solution of oil red O, saturated in 99 per cent isopropanol, with two parts of 1 per cent dextrin (to eliminate dye precipitation) and filtering just prior to use.

Method

1. Air dry sections and rinse in 60 per cent isopropanol;
2. Stain for 15 minutes in oil red O;

3. Differentiate in 60 per cent isopropanol until a delipidized control section appears colourless macroscopically;

4. Wash well in water and counterstain nuclei with Mayer's haemalum for three minutes;

5. Wash well in tap water, then distilled, and mount in glycerol—gelatin.

Results. Unsaturated hydrophobic lipids that are insoluble in the dye bath, and liquid at staining temperature (and mineral oils) stain red.

Method 5. Nile blue sulphate method for acidic lipids and neutral fats (Cain, 1947)

Fixation and sections. Cryostat sections post-fixed for one hour in calcium—formal; short-fixed frozen sections.

Preparation of reagent. Add $10 \, cm^3$ of 1 per cent H_2SO_4 to $200 \, cm^3$ 1 per cent Nile blue SO_4. Boil under reflux for four hours. Cool and filter. This solution should be at pH 2 so that non-lipid staining will be minimal.

Method

1. Air-dry sections on to slides;
2. Stain in the Nile blue SO_4 solution at $37°C$ for 30 minutes;
3. Differentiat sections in 1 per cent acetic acid for two minutes;
4. Wash well and mount in glycerol—gelatin;

Results. Unsaturated hydrophobic lipids — pink; free fatty acids — blue; phospholipids — blue.

Method 6. Acetone—Nile blue sulphate method for phospholipids (Dunnigan 1968)

Fixation and sections. As above.

Method

1. Mount sections on to slides;
2. Treat with 1N HCl for one hour to desoponify any calcium soaps;
3. Wash and dry sections;
4. Extract with acetone at $4°C$ for 20 minutes;
5. Dry sections and proceed as for the previous method.

Results. Phospholipids — blue.

Method 7. Phosphine 3R method for neutral fats (Popper 1944)

Fixation and sections. Calcium—formal fixed frozen sections; cryostat sections post-fixed in calcium—formal.

Method

1. Stain sections in 0.1 per cent aqueous phosphine 3R for 3 minutes;
2. Rinse briefly in water;
3. Mount in 90 per cent glycerine and examine in ultraviolet light;

Result: Cholesterol and glycerol esters fluoresce silvery-white.

8.2. Histochemical techniques

Once the bromine—Sudan black B screening test has confirmed that lipid is actually present in our specimen, we can proceed with the specific histochemical methods that are presently available to determine its identity. Bromine—acetone—Sudan black B and acetone—Nile blue SO_4 procedures will discriminate between fats and phospholipids. An oil red O-stained section viewed in partly polarized light is useful to indicate the type of lipid to be looking for with the specific procedures described below. Table 7 gives a protocol indicating the sequence to follow for the identification of an unknown lipid. In practice the investigator usually has some idea which type of lipid to expect, according to the nature of the tissue received and from hints about the clinical history of the lesion. Histochemical techniques exploit the unique chemical configurations of different lipid molecules so that they can be distinguished individually. The validity of about 60 such methods has been rigorously assessed against samples of pure lipids on filter paper (as in Fig. 2, p. 27) as well as with selected tissue sections. Only 30 can be regarded as sufficiently selective and reliable to be recommended herein. The methods of choice for each lipid class will be presented in detail below; other relevant techniques will receive only cursory and critical mention.

It must be reiterated that delipidized control sections should be included at all times to assess interference from non-lipid elements. It is also desirable to process 'positive' control material when available (see Chaper 6).

Free fatty acids

At the turn of the century, Benda (1900) showed that heavy metals bind to free fatty acids to form soaps, a principle that has provided the basis of a series of methods for the demonstration of fatty acids microscopically. Holczinger's (1959) method is the most sensitive and specific variant of a group of copper—soap procedures. Holczinger visualized the copper with rubeanic acid which produces an intense green-black pigment (see top left plate inside back cover).

An alternative colour reagent is Okomoto's dimethylaminobenzylidine—rhodanine which is less sensitive and slower to react than rubeanic acid. This factor however, proves to be an advantage because Holczinger's original method produces considerable overall staining, albeit authentic, due to the fatty acids within cell membranes, etc., whereas such background reaction is insignificant with the rhodanine reagent (Fig. 4); furthermore its red colour is compatible with a blue nuclear counterstain.

Calcium and iron deposits can base-exchange with copper and will therefore also react in these methods but are readily distinguished from fatty acids by comparison with a delipidized section, bearing in mind that calcium—formal fixation itself will already have converted fatty acids to their calcium soaps which would thus require to be desaponified before they can be extracted with acetone. The identity of

Table 7. *Scheme for the identification of tissue lipids*

Method	Reaction	Indication	Subsequent procedure
1.a. Bromine–Sudan black B	Blue-black	Lipid present	1b and c
b. Sudan black B (SBB)	Blue	Fats or myelin	2
c. Bromine–acetone–SBB	Grey	Phospholipids or lipofuscins	6 and 7
2.a. Nile blue SO$_4$	Pink	Neutral fats	3 and 4
	Blue	Fatty acids or phospholipids	5 and 6
b. Nile blue SO$_4$ in in polarized light	Birefringence	Crystalline cholesterol	3
c. Acetone–Nile blue SO$_4$	Blue	Phospholipids	6
3.a. PAN	Blue	Cholesterol	
b. Digitonin–PAN	Blue	Free cholesterol	
4. Calcium–lipase	Brown	Triglycerides	
5. Copper–rubeanic acid (or rhodanine)	Green ⎫ Red ⎭	Free fatty acids	
6.a. PAS	Magenta	Glycolipids or lipofuscins	7
b. Modified PAS	Magenta	Cerebrosides or gangliosides	6c and d
c. BH–periodate–Schiff	Pink ⎫	Gangliosides	
d. Thionin pH3	Pink ⎭		
e. DAH (or FeH)	Blue	Sphingomyelins or lecithins	6f and g
f. NaOH–DAH (or FeH)	Blue	Sphingomyelins	
g. Goldhydroxamic acid	Purple	Lecithins	
h. Acriflavine–DMAB	Scarlet ⎫	Sulphatides	
i. Toluidine blue–acetone	Brown ⎬		
j. High iron diamine	Purple ⎭		
7.a. Long Sudan black B	Grey ⎫		
b. PAS	Magenta ⎪		
c. Thionin pH3	Green ⎪		
d. Luxol fast blue	Blue ⎬ Ceroid–lipofuscins		
e. Long Ziehl Neelsen	Red ⎪		
f. Autofluorescence	Yellow ⎪		
g. Electron microscopy	Various forms ⎭		

pre-existing metal salt deposits can be verified by extracting them with either 1 per cent HCl (for calcium) or with 5 per cent oxalic acid (for iron salts).

Method 8. Copper–rubeanic acid method for free fatty acids (Holczinger 1959).

Fixation and sections. Cryostat sections post-fixed in calcium–formal; calcium–formal fixed frozen sections.

Method

1. Mount duplicate sections onto slides and allow them to dry;
2. Treat with 1N hydrochloric acid for one hour at room temperature to desaponify calcium soaps formed during fixation;
3. Wash well in distilled water and dry in air;
4. Extract one section with acetone at 4°C for 20 minutes and allow to dry;

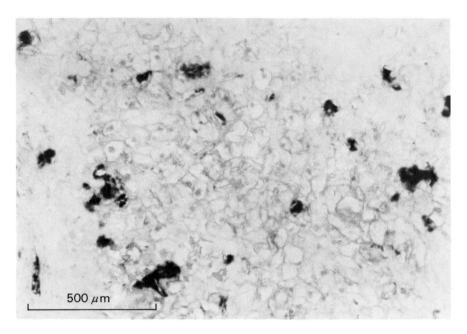

Fig. 4. Fatty liver with focal deposits of free fatty acids, stained dark red with the copper acetate–rhodanine method.

5. Immerse both sections in 0.005 per cent cupric acetate for three hours.

6. Wash twice for ten seconds with 0.1 per cent EDTA adjusted to pH 7.0 with NaOH;

7. Wash thoroughly in distilled water;

8. Treat sections for 10 minutes with 0.1 per cent rubeanic acid (dithio-oxamide) in 70 per cent ethanol;

9. Rinse in 70 per cent ethanol;

10. Counterstain nuclei with 2 per cent carmalum for 10 minutes;

11. Wash sections in water and mount them in glycerol–gelatin.

Results. Free fatty acids – dark green.

Notes

1. As an alternative to stage 8, copper soaps can be visualized by treating with 0.025 per cent dimethylaminobenzylidine–rhodanine in 70 per cent ethanol for 18 hours. Nuclei can be stained subsequently with Mayer's haemalum to provide a good contrast with the *red* reaction product of the fatty acids;

2. If unfixed cryostat sections are used, acetone extraction can be performed prior to fixation so that the stage of desaponification is unnecessary.

Cholesterol

The original Leibermann—Burchardt reaction for serum cholesterol was adapted for microscopy by Schultz (1924) using a surphuric—acetic acid mixture to give a transient blue-green colour with ferric alum-oxidized cholesterol in tissue sections. Later versions of this principle have improved but not overcome the problem of gaseous distortion of sections by such destructive reagents. Roussouw, Chase, Rath, and Engelbrecht (1976) avoided some of the technical difficulties associated with the classical procedure by using a 'premixed' reagent combining ferric chloride and acetic, sulphuric and phosphoric acids.

A perchloric acid—naphthoquinone(PAN) method, devised by Adams (1961) employs a totally different principle for cholesterol detection and is preferred for its sensitivity, specificity, and reasonable preservation of tissue sections. Perchloric acid is believed to condense cholesterol to cholesta-3,5-diene, which is then converted by 1,2-naphthoquinone to a blue pigment which persists for some hours.

Repurified cholesterol is entirely unreactive with all of the aforementioned methods. Neither the Schultz nor the PAN reagents will be effective unless the sections have been oxidized, either chemically or naturally after prolonged exposure to atmospheric oxygen. For the Schultz reaction, 2.5 per cent ferric alum is the most efficient oxidizing agent whereas 1 per cent ferric chloride is best for the PAN technique (Adams and Bayliss High 1980).

An interesting innovation in cholesterol histochemistry was an enzymatic technique introduced by Emeis, van Gent, and van Sabben (1977). Their two-stage procedure uses cholesterol esterase to hydrolyse esterified cholesterol to the free sterol; cholesterol is then oxidized by means of a cholesterol oxidase enzyme to release hydrogen peroxide which produces the usual insoluble brown polymer with diaminobenzidine(DAB) at the site of cholesterol. The free sterol can be visualized separately if the initial stage of ester hydrolysis is omitted, whereas the esters can be demonstrated selectively if free cholesterol has first been inactivated by the oxidase enzyme. Esterified cholesterol has also been detected enzymically using cholesterol esterase (Morii, Takigami, Kaneda, and Shikata 1982) in a method analogous to the calcium—lipase technique for triglycerides.

Although academically very interesting, the practical value of these enzyme methods for everyday light microscopy is somewhat limited by their cost and complexity, but both techniques have proved useful at the EM level (see Chapter 9).

Cholesterol can also be distinguished from its esters in the Schultz and PAN methods after digitonin precipitation. The cholesterol digitonide is insoluble in the acetone used to selectively extract cholesterol esters so that any subsequent staining can be attributed to the free sterol.

Method 9. Perchloric acid—naphthoquinone (PAN) method for cholesterol (Adams 1961)

Fixation and sections. Calcium—formal fixed frozen sections; cryostat sections post-fixed in calcium—formal.

Preparation of reagent

1:2 naphthoquinone-4-sulphonic acid	40 mg
Ethanol	20 cm^3
60 per cent perchloric acid	10 cm^3
40 per cent formaldehyde	1 cm^3
Distilled water	9 cm^3

Mix and use within 24 hours.

Method

1. Air dry sections on to slides;
2. Treat with 1 per cent ferric chloride for four hours to oxidize cholesterol;
3. Wash in several changes of distilled water;
4. Dry sections and paint them sparingly with reagent using a soft camel hair brush (which must be well washed afterwards). Heat slides on a warm surface at 70°C for one or two minutes, replenishing the reagent occasionally, until the colour develops. Take care not to char the section by overheating;
5. Mount section in a drop of perchloric acid.

Results. Cholesterol and its esters − blue. The colour is stable for several hours. Cholesterol can be distinguished from its esters by the following procedure.

Method 10. Digitonin−PAN method for free cholesterol

Fixation and sections. As above.

Method

1. Dry sections on to slides;
2. Precipitate free cholesterol with 0.5 per cent digitonin in 40 per cent ethanol for two hours at room temperature;
3. Extract cholesterol esters with acetone for one hour at room temperature;
4. Proceed with the PAN method described above;

Result. Free cholesterol − blue.

Method 11. Schultz method for cholesterol and its esters

Fixation and sections. As above.

Preparation of reagent. Add 1 cm^3 acetic acid dropwise, slowly to 1 cm^3 concentrated sulphuric acid with much agitation and cooling. The solution becomes hot and viscous but should remain colourless.

Method

1. Treat sections with 2.5 per cent ferric ammonium sulphate in 0.2 M acetate buffer at pH 3 for four hours at 37°C;
2. Wash three times during one hour in 0.2 M acetate buffer at pH 3;
3. Rinse in distilled water;

4. Treat with 5 per cent formalin in tap water for 10 minutes;

5. Drain and allow to dry;

6. Apply one or two drops of the reagent to the section and cover with a cover-slip.

Result. Cholesterol – a transient blue-green colour.

Unsaturated lipids

Occasionally it may be necessary to distinguish between saturated and unsaturated lipids. The reaction of ethylene groups (double bonds) in fatty-acid chains can be detected with osmium tetroxide, or perhaps more conveniently with Schiff's reagent after converting the double bonds to aldehydes with performic acid oxidation, or preferably with ultraviolet light, since performic acid treatment destroys tissue architecture and tends to act as a lipid solvent.

An alternative procedure introduced by Mukherji, Deb, and Sen (1960) involves bromination of the double bonds followed by silver nitrate to produce a silver bromide which can subsequently be reduced to visible metallic silver. However, the method of choice for demonstrating unsaturated lipids is the ultraviolet–Schiff technique devised by Belt and Hayes (1956).

Method 12. Ultraviolet—Schiff method for unsaturated lipids (Belt and Hayes 1956)

Fixation and sections. Unfixed cytostat sections if available; otherwise short-fixed frozen sections.

Method

1. Mount sections on to slides and expose to a source of ultraviolet light for two hours;

2. Treat with Schiff's reagent for 15 minutes, together with a non-irradiated control section to exclude non-lipid aldehydes;

3. Wash well in tap water, rinse in distilled, and mount sections in glycerol–gelatin.

Result. Unsaturated lipids—magenta.

Method 13. Osmium tetroxide method for unsaturated lipids

Fixation and sections. Preferably unfixed cryostat sections; otherwise use short-fixed frozen sections.

Method

1. Immerse sections in 1 per cent OsO_4 for one hour at room temperature;

2. Wash well in distilled water and mount sections in glycerol–gelatin.

Results. Unsaturated lipids—brown-back. Saturated lipids and free cholesterol—unstained.

Note. Osmium vapour is extremely toxic and this method should be carried out inside a fume cupboard.

Unsaturated lipids are demonstrated by osmium in this way because it is reduced to its black oxide by fatty-acid ethylene bonds. The main application of osmium in lipid histochemistry stems from an observation by Marchi in 1886 that if a highly polar electrolyte such as a dichromate salt is added to the osmium reagent, only degenerating myelin lipids (hydrophobic cholesterol esters) will be blackened, but normal myelin remains unstained. The classical Marchi method for the detection of degenerate myelin suffers a practical limitation in that whole tissue slices have to be osmicated for prolonged periods. Adams (1959) overcame this problem when he adapted the Marchi method for microscopy in his osmium tetroxide-α-naphthylamine (OTAN) procedure, which enables both normal and degenerating myelin to be appreciated in a single preparation. Adams based his method on the principle that osmium alone can penetrate hydrophibic fats such as the cholesterol esters encountered in demyelination, whereas the hydrophilic phospholipids of normal myelin are permeable to both osmium and the electrolyte, potassium chlorate. The latter, being highly polar, is therefore preferentially reduced by the unsaturated bonds within phospholipids, leaving the osmium in its colourless hexavalent form, subsequently to be chelated with α-naphthylamine to give the normal myelin an orange colour. One drawback to this method, the lengthy time of osmication, has been overcome by Negi and Stephens (1981) who formulated a briefer version of the OTAN technique. The method is highly specific and sensitive for detecting demyelination, but in view of the cost and toxicity of the reagents employed, a suitable alternative method is described later for the simultaneous demonstration of normal and degenerating myelin – namely the dichromate–acid haematein–oil red O technique.

Method 14. Osmium tetroxide-α-naphthylamine (OTAN) method (Adams 1959).

Fixation and sections. Cryostat sections post-fixed in calcium–formal: frozen sections of material fixed in calcium–formal.

Reagents

1. Osmium solution
   1 per cent osmium tetroxide   10 cm^3
   1 per cent potassium chlorate   30 cm^3
2. α-naphthylamine reagent. Add a few crystals of α-naphthylamine to 40 cm^3 distilled water at 40°C. Filter and use at 37°C.

Method

1. Treat sections for 18 hours in osmium–chlorate mixture in a well-stoppered container inside a fume cupboard;
2. Wash slides well for 10 minutes;
3. Treat sections with α-naphthylamine at 37°C for 20 minutes;
4. Wash well in distilled water;
5. Counterstain with 2 per cent alcian blue in 5 per cent acetic acid for 15 seconds;

6. Rinse in distilled water and mount sections in glycerol–gelatin.

Results. Cholesterol esters in degenerating myelin–black; phospholipids in normal myelin–orange.

Notes

1. Osmium vapour is toxic; α-naphthylamine may be contaminated with β-naphthylamine which is a carcinogen; therefore both these compounds should be handled with due care;

2. Negi and Stephens (1981) found that $2\frac{1}{2}$ hours in the osmium reagent is adequate for this method.

Triglycerides

Although triglycerides are the most abundant lipids in mammalian tissue and can be well stained with the conventional fat dyes, there was no specific method available for their identification until 1966 when Adams devised an enzymatic calcium–lipase technique whereby glycerol esters could be distinguished from esters of cholesterol and other fats. A pure lipase preparation, obtained from the pancreas of the pig, was used to hydrolyse the triglycerides in tissue sections, releasing their constituent fatty acids which, with calcium salts present in the incubation medium, are precipitated as calcium soaps. The calcium is then base-exchanged with lead and the lead soaps thus formed are finally visualized with ammonium sulphide – the brown lead sulphide indicating the original site of triglycerides. Pre-existing free fatty acids as well as calcium and iron deposits will be included in this reaction but can readily be distinguished from the true reaction by comparison with a control section that has been incubated without lipase.

Method 15. Calcium–lipase method for triglycerides (Adams, Abdulla, Bayliss and Weller 1966)

Fixation and sections. Cryostat sections post-fixed in calcium–formal; formal-fixed frozen sections.

Preparation of reagent

'Tris' buffer, 0.2 M, at pH 8.0	15 cm³
2 per cent calcium chloride	10 cm³
Distilled water	25 cm³
Porcine pancreatic lipase	50 mg

Warm solution to 37°C and filter before use.

Method

1. Incubate free-floating, or slide-mounted cryostat sections in the lipase medium at 37°C for three hours;

2. Wash sections well and mount them on to slides;

3. Together with a duplicate section that has not been subjected to enzymic hydrolysis, treat with 1 per cent lead nitrate for 15 minutes;

4. Wash very thoroughly in several changes of distilled water;

5. Immerse in 1 per cent ammonium sulphide for 20 seconds;

6. Wash well, counterstain with Mayer's haemalum for three minutes, and blue sections in tap water;

7. Rinse in distilled water and mount sections in glycerol–gelatin.

Results. Triglycerides – brown, nuclei –blue.

Notes

1. The reaction of free fatty acids and calcium deposits can be distinguished from a true reaction by inspecting the control section;

2. If fatty acids are present in large amounts, they can be selectively extracted with KOH–dioxane (Archibald and Orton 1970);

3. Lipolysis is confined to the water–fat interface; thus only the surface of large fat globules may be stained.

Phosphoglycerides

Lecithins and cephalins can be demonstrated selectively by a gold–hydroxamic acid method (Adams, Bayliss and Ibrahim 1963). This variant of the silver–hydroxamic acid technique (Gallyas 1963), itself a modification of the original ferric–hydroxamic acid technique (Adams and Davison 1959), depends on hydrolysis of the ester bond of phosphoglycerides with alkaline hydroxylamine to form hydroxamic acids. These then reduce silver nitrate to metallic silver which is finally converted to a stable purple reaction product by 'toning' with gold chloride. 'Fatty' hydrophobic esters are not included in this reaction because it takes place in aqueous reagents so that only hydrophilic esters (the phosphoglycerides) will be involved. Alkaline hydroxylamine is destructive to tissues, so they must be adequately fixed – not overlong, however, because lecithins are especially vulnerable to formalin fixation.

Method 16. Gold hydroxamic acid method for phosphoglycerides (Adams *et al.* 1963).

Fixation and sections. Cryostat sections fixed 18 hours in calcium–formal; short fixed frozens.

Preparation of reagents

1. Alkaline hydroxylamine
Hydroxylamine hydrochloride	2.5 g
Sodium hydroxide	6.0 g
Distilled water	100 cm^3

2. Silver nitrate solution

Silver nitrate	0.1 g
Ammonium nitrate	0.2 g
Distilled water	100 cm^3

Adjust to pH 7.8 with dilute sodium hydroxide.

Method

1. Treat free-floating frozen sections or slide-mounted cryostat sections (see note) with hydroxylamine solution for 20 minutes at room temperature;

2. Wash the sections thoroughly in three changes of distilled water, five minutes each change;

3. Immerse in the silver solution for two hours at room temperature;

4. Wash well, rinse in 1 per cent acetic acid and again in distilled water;

5. 'Tone' the sections for 10 minutes with 0.2 per cent yellow gold chloride;

6. Rinse in water;

7. Remove any unreduced silver with 5 per cent sodium thiosulphate for five minutes;

8 Wash well and mount free-floating sections on to slides;

9. Counterstain nuclei with 1 per cent methyl green (chloroform washed) for five minutes;

10. Mount in glycerol—gelatin or dehydrate; clear and mount in DPX.

Result. Phosphoglycerides—reddish-purple.

Note. Should cryostat sections become detached from their slides during alkaline hydrolysis, they can be processed free-floating thereafter, using a glass 'hockey stick' to transfer the sections gently from one solution to the next.

Plasmalogens

These phospholipids have an unsaturated ether bond which can be converted by mercuric chloride hydrolysis to an aldehyde group which is visualized with Schiff's reagent (Feulgen and Voit 1924). Since 'pseudoplasmal' aldehydes can be induced independently in double bonds by atmospheric oxidation, it is desirable to use fresh frozen sections promptly; otherwise interference from pseudoplasmals must be excluded by comparison with an unhydrolysed control section. Formalin fixation not only induces a strong pseudoplasmal reaction but also degrades the plasmalogens themselves, so should be avoided at all costs. Fortunately, unfixed cryostat sections are perfectly suitable for this method because the initial treatment with mercuric chloride effectively fixes the sections.

Method 17. Plasmal reaction for plasmalogen phospholipids (Hayes 1949)

Fixation and sections. Unfixed cryostat sections, stained as soon as possible after cutting.

Method

1. Air dry duplicate sections on to separate slides;
2. Hydrolyse one section in 2 per cent mercuric chloride for 10 minutes;
3. Wash thoroughly in three changes of distilled water;
4. Stain both sections with Schiff's reagent for 10 minutes;
5. Wash in running water for 10 minutes to develop the colour;
6. Counterstain nuclei with Mayer's haemalum for three minutes;
7. Wash well in tap water, rinse in distilled, and mount sections in glycerol–gelatin.

Results. Plasmalogens – magenta; nuclei – blue.

Notes

1. Any pseudoplasmal reaction would be present in the control section;
2. To avoid contaminating the Schiff's reagent with mercury, use a small quantity and discard after use.

Sphingomyelins

For over a century there has been a selection of well known histological techniques for staining myelin, based on the principle that metal salts act as mordants to increase the affinity of myelin for haematoxylin and other dyes in paraffin or celloidin sections. Baker (1946) devised a dichromate-acid haematein (DAH) technique for frozen sections with a precise schedule for fixation, chromation, and staining which is selective for the choline-containing lipids in myelin – namely lecithins and sphingomyelins. Baker's original procedure entailed chromation of whole blocks of tissue, which of course precludes the application of further techniques to the same specimen. The method has since been adapted for cryostat sections of unfixed material, thus reducing the processing time from days to hours.

Adams and Bayliss (1963) introduced a stage of alkaline hydrolysis prior to the dichromate–acid haematein sequence so that the reaction would be specific for sphingomyelin. Sodium hydroxide disrupts the ester bond of lecithins but leaves intact the amide linkage in the sphingomyelin molecule, so that the subsequent reaction will depict sphingomyelin alone. Tissues must be adequately fixed in order to withstand treatment with strong alkali but fortunately sphingolipids are not adversely affected by formalin.

Although the dichromate–acid haematein method is not the best myelin stain *per se,* it does have a useful application in combination with oil red O for the simultaneous demonstration of normal and degenerating myelin in frozen sections. A variety of neuro-pathological conditions exist where it may be necessary to discriminate between actively demyelinating lesions and old scars. The intracellular cholesterol esters derived from the degenerating myelin sheath are fatty and therefore stain well with oil red O, in contrast with the blue DAH reaction of residual, normal myelin (see plates top right and bottom left inside cover). The appearances of such Sudanophilic cholesterol esters is the hallmark of ongoing demyelination. The DAH–oil red O combination was originally proposed by Bourgeois and Hubbard

(1965) to distinguish between neutral fats and phospholipids, but their technique requires slight modifications if it is to be applied to neural tissues, to reduce the solubility of the cholesterol esters produced during myelin breakdown (Bayliss High 1982).

Method 18. Dichromate-acid haematein (DAH) method for choline-containing phospholipids (after Baker 1946)

Fixation and sections. Unfixed cryostat sections post-fixed in calcium—formal one hour; short fixed frozen sections.

Preparation of reagent

0.1 per cent haematoxylin	50 cm^3
1.0 per cent sodium periodate	1 cm^3

Heat to boiling, cool, and add 1 cm^3 glacial acetic acid.

Method

1. Treat sections with 5 per cent potassium dichromate containing 1 per cent calcium chloride for 18 hours at room temperature, followed by a further two hours at 60°C;
2. Wash well in several changes of distilled water for 30 minutes;
3. Stain with the acid haematein solution for two hours at 37°C;
4. Wash well and differentiate in an aqueous solution of 0.25 per cent sodium tetraborate and 0.25 per cent potassium ferricyanide for two hours at 37°C;
5. Wash well in distilled water;
6. Counterstain nuclei with 1 per cent methyl green;
7. Wash well and mount sections in glycerol—gelatin.

Results. Lecithin and sphingomyelin — blue; nuclei — green.

Notes

1. A control section, delipidized in chloroform:methanol (2:1), should be included to assess the extent of any non-lipid staining;
2. Several batches of haematoxylin have proved to be unsatisfactory for this method; that supplied by Sigma chemical Co. Ltd is recommended;
3. This method can be followed by oil red O-staining to demonstrate normal and degenerating myelin simultaneously, blue and red, respectively (see bottom left plate inside back cover).

Method 19. Sodium hydroxide — dichromate acid haematein method for sphingomyelin (Adams and Bayliss 1963)

Fixation and sections. Cryostat or frozen sections, mounted on to chrome—gelatin slides and fixed for at least 24 hours in calcium—formal.

Method

1. Treat sections with 2 M sodium hydroxide for one hour at 37°C;
2. Wash slides gently but briefly in a large volume of water;

3. Rinse in 1 per cent acetic acid for five seconds;

4. Remount sections that might have become detached from their slides on to chrome gelatine slides and proceed as for the dichromate–acid haematein method described above.

Result. Sphingomyelin – blue.

Another version of the metal–haematoxylin principle was introduced by Elleder and Lojda (1973). This ferric–haematoxylin (FeH)method is simpler, quicker and apparently more sensitive than the DAH method and after alkaline hydrolysis has been especially useful for detecting sphingomyelin in Niemann Pick's disease. Although FeH may be the method of choice for sphingomyelin, staining is satisfactory only after sections have been defatted with acetone, thereby precluding the application of this technique in combination with oil red O, as described above for the DAH method. Furthermore, the FeH procedure causes nuclei to be stained in the same blue as lipid; in some instances this may be disadvantageous.

Method 20. Ferric haematoxylin (FeH) method for phospholipids (Elleder and Lojda 1973)

Fixation and sections. Ideally unfixed cryostat sections, otherwise short-fixed frozen sections.

Preparation of reagent

Solution a.	Distilled water	298 cm^3
	Concentrated hydrochloric acid	2 cm^3
	FeCl$_3$ · 6H$_2$O	2.5 g
	FeSO$_4$ · 7H$_2$O	4.5 g
Solution b.	Distilled water	100 cm^3
	Haematoxylin	1 g
	Dissolve by gentle heating.	

Working solution. Mix three parts solution a with one part solution b and use within one hour.

Method

1. Treat duplicate air-dried sections (a and b) as follows.
   a. Extract in chloroform:methanol (2:1) for one hour at room temperature;
   b. Extract with acetone at 4°C for 15 minutes.
2. Fix both sections in calcium–formal for 30 minutes;
3. Rinse in distilled water;
4. Stain in ferric haematoxylin for seven minutes;
5. Wash in distilled water;
6. Dip several times in 0.2 per cent HCl;
7. Wash in tap water;
8. Dehydrate in acetone, clear in xylene, and mount in DPX.

Results. Phospholipids – blue.

Method 21. Sodium hydroxide—ferric haematoxylin method for sphingomyelin

Fixation and sections. As above.

Method

1. Treat section with 2 M sodium hydroxide for one hour at room temperature;
2. Wash gently but briefly in a large volume of water;
3. Rinse in 1 per cent acetic acid for five seconds;
4. Remount section if it has become detached during hydrolysis or washing and proceed with the ferric haematoxylin method as described above.

Results. Sphingomyelin — blue.

Note. Should plasmalogens interfere, they can be excluded by treating sections with 1 per cent mercuric chloride in 1 per cent hydrochloric acid for 10 minutes prior to alkaline hydrolysis.

Cerebrosides

These glycolipids on account of their hexose moieties are stained by the periodic acid—Schiff (PAS) method, originally designed to stain mucopolysaccharides, when periodic acid converts the 1,2-glycol group in the hexose molecule to Schiff—stainable aldehyde. Since other chemical configurations would also react in this way, Adams and Bayliss (1963) devised a sequence of blockades whereby interfering groups can be suppressed so that PAS reactivity is attributable solely to hexoses.

The series of blockades consists first of deamination with chloramine T to convert amino groups to carbonyls; then lipid ethylene bonds are oxidized to aldehydes by performic acid so that, together with any pre-existing aldehyde groups, they can be blocked with dinitrophenyl hydrazine. The PAS method is then performed to stain hexoses, their 1,2-glycol groups having been unaffected by the foregoing treatment. The reaction will include not only the hexose groups of glycolipids but also those in glucosaminoglycans so that the presence of glycolipid can only be ascertained by comparison with a delipidized control section. Such 'difference' methods are not ideal but in practice the results are usually clear-cut, with one exception — namely the liver in Gaucher's disease. If one wishes to demonstrate the stored cerebroside in this metabolic disorder, glycogen in the liver will provide such an intense non-lipid reaction that traces of cerebroside may be overlooked. Consequently, a control section treated with diastase should be included, to exclude the reaction of interfering glycogen.

Method 22.Modified PAS reaction for cerebrosides (Adams and Bayliss 1963)

Fixation and sections. Cryostat sections post-fixed in calcium—formal; calcium—formal fixed frozen sections.

Preparation of reagent

Performic acid	
98 per cent formic acid	45 cm^3
100 vols hydrogen peroxide	4.5 cm^3
Concentrated sulphuric acid	0.5 cm^3

Prepare an hour before use and stir occasionally with a glass rod to release gas bubbles form the solution. The reagent is extremely destructive to skin and clothing so beware of splashes and vapour.

Method

1. Mount duplicate sections on to separate slides and extract one of these with chloroform: methanol (2:1 v/v) for one hour at room temperature;

2. Deaminate both sections in 10 per cent aqueous chloramine T for one hour at 37°C;

3. Wash slides vigorously and as rapidly as possible, one at a time, in a large volume of water before transferring them immediately to the performic acid solution. This washing stage must be swift, yet thorough, to avoid swelling and detachment of the sections from their slides;

4. Immerse in performic acid for 10 minutes;

5. Wash well in several changes of distilled water;

6. Treat with a filtered solution of 2:4 dinitrophenyl hydrazine, saturated in 1 M hydrochloric acid, at 4°C for two hours;

7. Wash well in distilled water;

8. Treat with 1 per cent periodic acid for 10 minutes;

9. Wash in distilled water;

10. Stain in Schiff's reagent for 15 minutes;

11. Wash in tap water for 15 minutes to allow the colour to develop;

12. Counterstain nuclei with Mayer's haemalum;

13. Wash well in tap water, rinse in distilled and mount sections in glycerol–gelatin.

Results. Cerebrosides – magenta, indicated by the difference in staining between test and control sections.

Note. Protein-bound ganglioside may also stain but can be distinguised from cerebroside by digesting a control section with the enzyme neuraminidase as described in the section on gangliosides below.

Sulphatides

These sulphuric-acid esters of cerebroside do not stain with the PAS method because their reactive 1,2-glycol groups have been blocked by esterification. However, sulphatides are the only lipids that are sufficiently acidic to induce a metachromatic shift in a variety of basic aniline dyes which have been used to demonstrate the abnormal lipid deposits in sulphatide storage disease, consequently referred to as metachromatic leucodystrophy. With cresyl violet, for example, sulphatide appears orange in contrast to the orthochromatic colour of other, less acidic myelin lipids. Trivial metachromasia may be difficult to appreciate against a strong purple background but distinction can be improved if the stained preparation is viewed in polarized light so that the sulphatide displays a green dichroism. Feyrter's thionin method is also hampered by an overpowering background colour but if toluidine blue is followed by acetone extraction (Bodian and Lake 1963) then sulphatide is selectively and distinctly stained metachromatically reddish-brown.

A different staining principle underlies Holländer's (1963) adaptation of an acriflavine method originally designed to stain sulphated mucins. Holländer's (1963) adaptation of an acriflavine method originally designed to stain sulphated mucins. Holländer's reagent is sufficiently acidic to allow sulphatide, exclusively among lipids, to be selectively stained. Interference from non-lipid substances such as heparin within mast cell granules, is readily distinguished by its persistence in a delipidised control section. The stained acriflavine–sulphatide complex can be viewed in ultraviolet light in which it emits an orange fluorescence; alternatively the lipid–dye complex can be converted to a stable, visible, scarlet pigment with *p*-dimethylaminobenzaldehyde (DMAB) giving excellent localization of sulphatide against a colourless ground.

Another method that can be usefully adapted to frozen sections for staining sulphatide is the high iron diamine method (Spicer 1965) used conventionally to stain sulphated mucins in paraffin sections. Among lipids the staining reaction is confined to sulphatide (see Fig. 5) and the possibility of interference from cross-reacting mucins can be assessed in a delipidized control section.

Method 23. Acriflavine–DMAB method for sulphatide (Holländer 1963)

Fixation and sections. Post-fixed cryostat sections; calcium–formal frozen sections.

Preparation of reagents

   a. Acriflavine stock solution
Acriflavine	100 mg
Distilled water at 80°C	20 cm^3

      Store in darkness at 4°C.
   b. Acriflavine working solution
0.1 M citrate–HCl buffer pH 2.5	99 cm^3
Stock acriflavine solution	1 cm^3
   c. DMAB solution
p-dimethylaminobenzaldehyde	0.6 g
20 per cent hydrochloric acid	30 cm^3
Isopropanol	70 cm^3

Method

   1. Stain sections for six minutes in the acriflavine working solution;
   2. Differentiate for one minute in two changes of 70 per cent isopropanol;

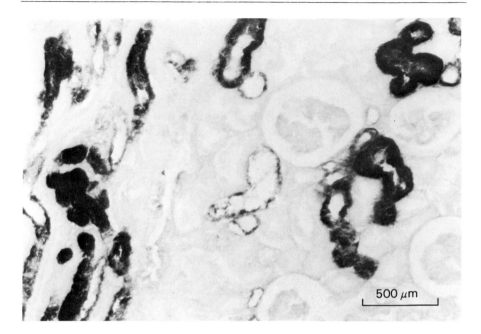

Fig. 5. Sulphatide in renal tubules from a patient with metachromatic leucodystropy. High iron diamine–Alcian blue.

3. Treat with DMAB reagent for 30–45 seconds;
4. Rinse in distilled water for 2–3 minutes;
5. Counterstain with Mayer's haemalum for three minutes;
6. Blue in tap water, rinse in distilled, and mount sections in glycerol–gelatin.

Results. Sulphatide – red. Nuclei – blue.

Notes

1. Mast-cell granules appear red but resist extraction in chloroform: methanol;
2. Alternative after stage 2. Dehydrate sections in isopropanol, clear in xylene, and mount in a fluorescence-free medium and view sections in ultraviolet light.

Result. Sulphatide – orange in U.V. light against a green ground.

Method 24. Toluidine blue – acetone method for sulphatide (Bodian and Lake 1963)

Fixation and sections. Post-fixed cryostat sections: calcium–formal fixed frozen sections.

Preparation of reagent.
0.01 per cent toluidine blue in phosphate–citrate buffer at pH 4.7. The buffer comprises

0.2 M disodium hydrogen phosphate   96 cm^3
0.1 M citric acid   104 cm^3

Method

1. Stain sections for 16—18 hours in buffered toluidine blue;
2. Wash in water;
3. Treat with acetone for five minutes;
4. Clear sections in xylene and mount them in DPX.

Result. Sulphatide—reddish-brown.

Method 25. High iron diamine method for sulphatide (after Spicer 1965)

Fixation and sections. Frozen sections of calcium—formal fixed material or cryostat sections post-fixed in calcium—formal.

Preparation of reagent

N,N-dimethyl-meta-phenylene diamine dihydrochloride	120 mg
N,N-dimethyl-para-phenylene diamine dihydrochloride	20 mg
Distilled water	50 cm^3
Ferric chloride (60 per cent, British Drug Houses)	1.4 cm^3

Dissolve the two diamine salts simultaneously in the water; add the ferric chloride and mix. Leave the solution to stand for 30 minutes to allow the ferric chloride to oxidize the diamine salts to a purple cationic chromogen.

Method

1. Treat sections for 18 hours at room temperature with the reagent;
2. Wash well in running water;
3. Counterstain if desired with 1 per cent alcian blue in 3 per cent acetic acid for five minutes;
4. Wash in water, dehydrate, clear, and mount in DPX.

Results. Sulphatide —purple; background — pale blue.

Notes

1. The two diamine salts are potentially toxic and should be handled with extreme caution;
2. Possible cross-reaction from sulphomucins can be excluded by comparison with a delipidized control section.

Gangliosides

Gangliosides can be distinguished histochemically from other glycolipids on account of their constituent neuraminic acids and their acyl derivatives — the sialic acids. Early methods attempting to exploit this chemical configuration were unsatisfactory until Ravetto (1964) successfully adapted the Svennerholm—Bial chromatography reagent to stain gangliosides in tissue sections. However, the colour obtained in this way is unacceptably pale and the mehod is not pleasant to perform. The reagent, consisting of copper sulphate and orcinol dissolved in concentrated HCl, has to be applied in spray form so that the supposedly water-soluble gangliosides are not washed out of the section. In practice however, the intraneuronal accumulations of ganglioside in the storage disorders involving this lipid happen to be protein-bound and therefore are not only insoluble in water but also largely survive processing for paraffin embedding. As outlined in Chapter 9, such lipids have been demonstrated in paraffin sections using an immunocytochemical technique. Since the ganglioside storage disorders provide the sole occasion for which ganglioside requires to be demonstrated histochemically, the water-solubility of free ganglioside is scarcely relevant microscopically.

In fact the conventional PAS method has proved to be the most practical and rewarding approach to ganglioside detection. Although the PAS reaction includes other glycolipids and of course non-lipid mucosubstances, Roberts (1977) devised a modification whereby Schiff-staining could be confined to glycoproteins containing sialic acids. Sialo groups are oxidized much more rapidly than other sugar glycosides and so by drastically reducing the concentration of the oxidizing agent (0.01 per cent metaperiodate in place of 1 per cent periodic acid) the reaction would be confined to sialomucins.

It was suggested to me that the same manoeuvre might be relevant to sialo-lipids (S.J.A. Buk, personal communication) and indeed the method does convincingly stain gangliosides within neurones in Tay Sach's disease, if the strength of the periodate solution is slightly increased (see bottom right plate inside cover).

Gangliosides, on account of their carboxyl groups, can also be stained by the basic aniline dyes such as thionin, cresyl violet, and Nile blue SO_4 but these dyes are by no means selective and will also colour phospholipids due to their phosphoric-acid residues and sulphatide on account of its sulphate radicle. A rose-purple metachromasia produced immediately with Feyrter's thionin method (see method 28) is a useful indicator of sialic-acid residues.

Method 26. Borohydride–periodate–Schiff (BHPS) method for gangliosides (after Roberts 1977)

Fixation and sections. Cryostat sections post-fixed in calcium–formal; frozen sections of fixed tissue.

Method

1. Destroy existing carbonyls by reducing sections with 0.1 M (0.38 per cent) sodium borohydride in 1 per cent disodium hydrogen phosphate for one hour at room temperature;

2. Wash thoroughly in distilled water;

3. Oxidize with 1.2 mM (0.03 per cent) sodium metaperiodate for 30 minutes at room temperature;

4. Wash twice for five minutes each time in distilled water;

5. Stain with Schiff's reagent for 10 minutes;

6. Wash well in tap water and then in distilled;

7. Counterstain with Mayer's haemalum for five minutes;

8. Blue in tap water, rinse in distilled, and mount sections in glycerol–gelatin.

Results. Gangliosides in Tay Sach's disease — pink; nuclei — blue.

Notes

1. A chloroform: methanol extracted section should be used for comparison to exclude interference from non-lipid sialomucins;

2. The reaction can be verified by comparison with a sialidase-digested control section. Pretreat sections for 18 hours at 37°C with sialidase (neuraminidase), ex vibrio cholerae, available from Burroughs Wellcome and diluted 1:4 with 0.1 M acetate buffer at pH 5.5, containing 0.9 mM calcium chloride.

3. Alternatively, use 0.1 M sulphuric acid for two hours to hydrolyse sialic acid residues.

Neither enzyme digestion nor acid hydrolysis is entirely satisfactory with formalin-fixed tissues, although the staining reaction is much reduced thereby.

Lipofuscins

Pathologically the most significant of these lipopigments are the intraneuronal deposits of ceroid—lipofuscin found in the Batten group of metabolic storage disorders. The stored granules are derived from lysosomal lipids that have undergone varying degrees of peroxidation and polymerization, becoming increasingly insoluble in organic solvents, more acid-fast (i.e. strongly stained with Ziehl Neelsen) and correspondingly less Sudanophilic. Intraneuronal deposits of such compounds appear grey with Sudan black B; they are PAS-positive and are fluorescent in ultraviolet light. Some types of lipofuscin can be stained with luxol fast blue but electron-optically their ultrastructural features provide the most convincing diagnostic evidence in the group of ceroid lipofuscinoses (see Fig. 6). The three main variants of Batten's disease, graded according to the age of onset, are tabulated below (Table 8) to indicate the variable staining and ultrastructural properties of their associated lipopigments.

Methods for ceroid—lipofuscin deposits in Batten's disease

Fixation and sections. Unfixed cryostat sections cut at $7\,\mu$.

Method 27. Long Sudan black B

Method. As described in Method 2 but leave overnight in the dye solution.

Results. Lipofusins — grey.

Method 28. Periodic acid—Schiff (PAS)

Method. Apply the standard technique to duplicate sections, one of which has been coated with celloidin (0.25 per cent celloidin in ethanol) to retain water-soluble compounds.

Results. Some lipofuscins — magenta.

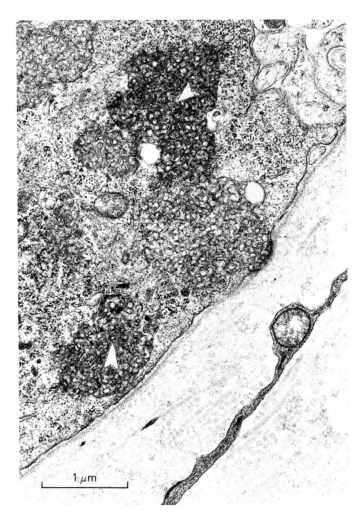

Fig. 6. Electron micrograph showing 'curvilinear bodies' (arrowed) = intraneuronal accumulations of ceroid–lipofuscin in the appendix from a case of Batten's disease.

Table 8. *Histochemical features of Batten's disease (after Lake 1976)*

Variant	SBB	PAS	LFB	AF	EM
Infantile (Santavuori)	++	++	—	++	Membrane-bound bodies
Late infantile (Bielschowsky–Jansky)	+	—	+	+	Curvilinear bodies
Juvenile (Speilmeyer–Vogt or Sjogren)	++	++	++	+++	Fingerprint bodies

Method 29. Thionin pH 3

Preparation of reagent

Thionin	1 g
Tartaric acid	0.5 g
Distilled water	100 cm^3

Method

1. Cover sections with the dye and apply a coverslip;
2. Seal with paraffin wax and examine microscopically after five minutes.

Results. Lipofuscins — green. (Gangliosides and cerebrosides would appear rose-pink).

Method 30. Luxol fast blue

Method

1. Rinse sections in 95 per cent ethanol;
2. Stain for 16–18 hours at 60°C in 0.1 per cent Luxol fast blue in 95 per cent ethanol;
3. Rinse in 70 per cent ethanol;
4. Wash in tap water.
5. Differentiate in 0.05 per cent lithium carbonate until no further dye comes away (overdifferentiation is impossible);
6. Wash thoroughly in water;
7. Counterstain with 1 per cent aqueous neutral red;
8. Rinse, dehydrate, clear, and mount in DPX.

Results. Some ceroid—lipofuscins — blue; nuclei — red.

Method 31. Long Ziehl—Neelsen method for lipofuscins

Preparation of reagent

Basic fuchsin	1 g
Phenol	0.5 g
Absolute ethanol	10 cm^3
Distilled water	100 cm^3

Method

1. Stain sections in carbol fuchsin for three hours at 60°C;
2. Wash well in running water;
3. Differentiate in 1 per cent acid alcohol until background is clear;
4. Wash well in tap water;
5. Counterstain in 2 per cent methylene blue for one minute;
6. Wash well, dehydrate, and clear. Mount in DPX.

Results. Lipofuscins — magenta; nuclei — blue.

Method 32. Autofluorescence

Method

1. Rinse cryostat sections in xylene and mount them in DPX;
2. Examine by fluorescence microscopy with dark-ground illumination, using an excitation filter at 300–70 nm (UG5) and a barrier filter of 410 nm.

Results. Ceroid–lipofuscins–yellow fluorescence.

Recent advances in lipid histochemistry

9.1. Histochemical combination techniques

During the past few decades, histologists have successfully devised a number of staining procedures whereby two or more tissue elements can be depicted simultaneously in contrasting tones for microscopy. Unlike staining procedures, histochemical techniques are less likely to be compatible when applied in sequence, because the first method may inactivate or extract the object of a second reaction. There are of course notable exceptions, such as the high iron diamine—Alcian blue combination for the differential staining of sulphated and carboxylated mucins. Lipid methods, requiring frozen sections, pose additional problems. However, the Table 9 lists some examples of 'dual' techniques that have proved useful either to demonstrate two or more lipids in a single preparation, or to depict fats and enzyme activity together. Such combinations can be far more informative than the same techniques applied separately to serial sections. A combined reaction in a single section will, for example, reveal more accurately the relationship between enzymically active phagocytic cells and associated lipid deposits.

9.2. Electron histochemistry of lipids

Understandably, lipids have received scant attention at the EM level largely because the few reaction products that would remain insoluble during processing are rarely electron-dense. Although osmic acid preserves most phospholipids, their individual identities are not thereby revealed. Osmicated fats are generally soluble in the processing fluids although retention can be improved by adding imidazole (Angermüller and Fahimi 1982). A method has been described for cholesterol as its osmiophilic digitonide (Ökrös 1968) and the enzymic cholesterol method described by Emeis *et al.* (1977) has been successfully adapted for the electron microscope (Martin Jones and Miyai 1981). Phosphoglycerides have been localized at the silver stage in the gold hydroxamic acid method described previously; triglycerides are also electron-dense after the calcium—lipase procedure and likewise an enzyme digestion method enables cholesterol esters to be visualized electron-optically (Morii *et al.* 1982).

Undoubtedly, the most rewarding application of electron microscopy to lipid histochemistry has been in the study and diagnosis of the metabolic storage

Table 9. *Histochemical combination techniques*

Method	Components demonstrated	Colours
1. Dichromate-acid haematein–	Phospholipids (normal myelin)	Blue
oil red O–	Fats (degenerating myelin)	Red
Methyl green	Nuclei	Green
2. Copper acetate–rhodanine–	Free fatty acids	Red
Sudan black B–	Neutral fats	Blue
Polarized light	Crystalline cholesterol	Birefringent
3. PAS–	Glycolipids	Pink
Sudan black B	Neutral fats and myelin	Blue
4. Acid phosphatase (azo dye method)	Lysosomes	Red
Sudan black B	Neutral fats and myelin	Blue
5. β-galactosidase–	Lysosomes	Turquoise
Oil red O	Fats	Red
6. Acid esterase–	Lysosomes	Orange
Sudan black B–	Neutral fats and myelin	Blue
Methyl green	Nuclei	Green
7. Succinate dehydrogenase–	Mitochondria	Blue
Oil red O–	Fats	Red
Methyl green	Nuclei	Green

disorders such as Batten's disease and the gangliosidoses, when the neuronal storage granules exhibit distinctive ultrastructural patterns (see Fig. 6, p. 56) and these may provide the most convincing diagnositc evidence in such conditions.

9.3. Immunohistochemistry of lipids.

Immunoperoxidase and immunofluorescence techniques have revolutionized almost every aspect of histochemistry in recent years by allowing individual compounds and their subgroups to be identified more precisely than with most of our conventional staining techniques. Lipids have received little attention from the immunologists, probably because most of the major lipid classes are non-antigenic and the few that are antigenic pose considerable technical difficulties in immunohistochemistry. Although cardiolipin, a major factor of the mitochrondrial membrane, is the factor in beef-heart extract that is responsible for the serological test for syphilis, this reaction has not been exploited histochemically.

The first real interest in the immunogenicity of lipids stemmed from the recognition of a glycosphingolipid antigen in cell membranes. This group of lipids includes some of the blood-group antigens and has also been detected in tumour cells. However, this field seems not to have been developed microscopically as yet, with the exception of galactocerebroside in myelin and the gangliosides, which are the characteristic glycolipids of neural tissue – the counterparts of globosides in other tissues. Antisera have been raised against GM_1 ganglioside for the immunohistochemical localization of gangliosides in the cerebellar cortex (de Baeque, Johnson, Naiki, Schwarting, and Marcus 1976). Such antibodies were shown to be directed against the trihexosyl moiety of the ganglioside molecule, which acts as a hapten (i.e. requiring to be conjugated with a protein carrier before it can elicit an immune response). The antibodies raised were of the IgG and IgM classes. GM_2 antiserum raised in the rabbit to human ganglioside produced a granular reaction in the neurones of a patient with Tay Sach's disease, some of the reaction persisting even in paraffin-embedded material (Schwerer, Lassman, and Bernheimer 1982).

Apart from the special case of the glycolipids, immunohistochemistry is unlikely to have further applications in this field since other lipids are non-immunogenic. Consequently, the conventional histochemical techniques described in detail in this handbook presently constitute the best means of localizing and identifying lipids microscopically.

Appendix: list of methods

References

Adams, C.W.M. (1959). A histochemical method for the simultaneous demonstration of normal and degenerating myelin. *J. Path. Bact.* **77**, 648–50.
────── (1961). A perchloric acid naphthoquinone method for the histochemical demonstration of cholesterol. *Nature, London* **193**, 331–2.
────── (Ed.) (1965). Neurohistochemistry. Elsevier, Amsterdam.
──────, Abdulla, Y.H., Bayliss, O.B., and Weller, R.O. (1966). Histochemical detection of triglyceride esters with specific lipases and a calcium–lead sulphide technique. *J. Histochem. Cytochem.* **14**, 385–95.
────── and Bayliss, O.B. (1963). Histochemical observations on the localisation and origin of sphingomyelin, cerebroside and cholesterol in normal and atherosclerotic human artery. *J. Path. Bact.* **85**, 113–19.
────── and ────── (1980). Preliminary oxidation in histochemical staining methods for cholesterol. *J. Microscopy* **119**, 427–30.
──────, ──────, Baker, R.W.R., Abdulla, Y.H., and Hunter-Craig, C.J. (1974). Lipid deposits in ageing human arteries, tendons and fascia. *Atherosclerosis* **19**, 429–40.
──────, ──────, and Ibrahim, M.Z.M. (1963). Modifications to histochemical methods for phosphoglyceride and cerebroside. *J. Histochem. Cytochem.* **11**, 560–1.
────── and Davison, A.N. (1959). The histochemical identification of myelin phosphoglycerides by their ferric hydroxamates. *J. Neurochem.* **3**, 347–351.
Angermüller, S. and Fahimi, H.D. (1982). Imidazole-buffered osmium tetroxide: an excellent stain for visualization of lipids in transmission electronmicroscopy. *Histochem. J.* **14**, 823–5.
Archibald, R.W.R. and Orton, C.C. (1970). Specific identification of free and esterified fatty acids in tissue sections. *Histochem. J.* **2**, 411–17.
Baker, J.R. (1945). Structure and chemical composition of the Golgi element. *Quart. J. Microsc. Sci.* **85**, 1–71.
────── (1946). The histochemical recognition of lipine. *Quart. J. Microsc. Sci.* **87**, 441–70.
Bancroft, J.D. and Stevens, A. (Eds.) (1982). *Theory and practice of histological techniques*, 2nd edn. Churchill Livingstone, Edinburgh.
Bayliss, O.B. (1976). The giant cell in cholesterol resporption. *Brit. J. exp. Pathol.* **57**, 610–18.
────── (1981). The histochemical versatility of Sudan black B. *Acta Histochem. Suppl., Band. XXIV*, **5**, 247–55.
────── (1982). Histochemical combinations. *Proc. Roy. microsc. Soc.* **17**, 132.
────── and Adams, C.W.M. (1972). Bromine Sudan Black (BSB). A general stain for tissue lipids including free cholesterol. *Histochem. J.* **4**, 505–15.
Belt, W.D. and Hayes, E.R. (1956). An ultra violet–Schiff reaction for unsaturated lipids. *Stain Technol.* **31**, 117–22.

Benda, C. (1900). Eine makro- und mikrochemisch Reaction der fett Gewebsnekrose. *Virchow's Arch.* **161**, 194–8.

Bodian, M. and Lake B.D. (1963). The rectal approach to neuropathology. *Br. J. Surg.* **50**, 702–14.

Bourgeois, C. and Hubbard, B. (1965). A method for simultaneous demonstration of choline containing phospholipids and neutral lipids in tissue sections. *J. Histochem. Cytochem.* **13**, 571–8.

Cain, A.J. (1947). Use of Nile Blue in the examination of lipoids. *Quart. J. microsc. Sci.* **88**, 383–92.

de Baecque, C., Johnson, A.B., Naiki, M., Schwarting, G., and Marcus, D.M. (1976). Ganglioside localisation in cerebellar cortex: an immunoperoxidase study with antibody to GM$_1$ ganglioside, *Brain Res.* **114** (11), 7–122.

de Groot, P.A. and den Hartog Jager, W.A. (1980). A storage product in Pick's presenile dementia. *Abstracts VIth Internat. Histochem. and Cytochem. Congress Roy. micro. Soc.,* p. 151.

den Hartog Jager, W.A. (1969). Sphingomyelin in Lewy inclusion bodies in Parkinson's disease. *Arch. Neurol.* **21**, 615–19.

Diezel, P.B. (1957). Histochemical studies of primary lipidoses. In *Cerebral lipidoses* (ed. J.N. Cumings), pp. 11–29. Blackwell, Oxford.

Dunnigan, M.G. (1968). The use of Nile Blue Sulphate in the histochemical identification of phospholipids. *Stain Technol.* **43**, 249–56.

Elleder, M. and Lojda, Z. (1973). New, rapid, simple and selective method for the demonstration of phospholipids. *Histochemie* **36**, 149–66.

Emeis, J.J., van Gent, C.M., and van Sabben, C.M. (1977). An enzymatic method for the histochemical localisation of free and esterified cholesterol separately. *Histochem. J.* **9**, 197–204.

Feulgen, R. and Voit, K. (1924). Ueber einen Weitverbreiteten festen Aldehyd seine Enstehung aus einer vorstufe, sein Mikrochemischer Nachweis und die Wege zuseiner praparativen Darstellung. *Pflüg. Arch.* **206**, 389–410.

Gallyas, F. (1963). The histochemical identification of phosphoglycerides in myelin. *J. Neurochem.* **10**, 125–6.

Hayes, E.R. (1949). A rigorous redefinition of the plasmal reaction. *Stain Technol.* **24**, 19–23.

Holczinger, L. (1959). Histochemischer Nachweis freier Fättsauren. *Acta Histochem.* **8**, 167–75.

Holländer, H. (1963). A staining method for cerebroside–sulfuric esters in brain tissue. *J. Histochem. Cytochem.* **11**, 118–19.

Lake, B.D. (1976). The differential diagnosis of the various forms of Batten disease by rectal biopsy. *Birth defects* **XII(3)**, 455–62.

Lillie, R.D. and Ashburn, L.L. (1943). Supersaturated solutions of fat stains in dilute isopropanol for demonstration of acute fatty degeneration not shown by Herxheimer's technic. *Arch. Pathol.* **36**, 432–5.

Lison, L. (1936). *Histochemie animale,* 1st edn. Gautier: Villars, Paris.

Lovern, J.A. (1955). *The chemistry of lipids of biochemical significance.* Methuen, London.

Marchi, V. (1886). Sulle degenerazioni consecutive all'estirpazione totale e parziale del cervelletto. *Riv. Sper. Freniat.* **12**, 50–6.

Martin Jones, H. and Miyai, K. (1981). Ultrastructural localisation of cholesterol by enzyme histochemistry. *Histochem. J.* **13**, 1017–28.

Morii, S., Takigami, S., Kaneda, Y., and Shikata, N. (1982). Ultracytochemical analysis of cytoplasmic lipids by enzymic digestive methods. *Acta Histochem. Cytochem.* **15**, 185–91.

Mukherji, M., Deb, C., and Sen, P.B. (1960). Histochemical demonstration of unsaturated lipids by bromine silver method. *J. Histochem. Cytochem.* **8**, 189–94.

Negi, D.S. and Stephens, R.J. (1981). Rapid OTAN method for localizing unsaturated lipids in lung tissue sections. *Stain Technol.* **56**, 177–80.

Ökrös, I. (1968). Digitonin reaction in electron microscopy. *Histochemie* **13**, 91–6.

Pearse, A.G.E. (Ed.) (1968). *Histochemistry: theoretical and applied*, 3rd edn., Vol. 1, Churchill, London.

Popper, H. (1944). Distribution of Vitamin A in tissue as visualised by fluorescence microscopy. *Physiolog. Rev.* **24**, 205–24.

Ravetto, C. (1964). Histochemical identification of sialic (neuraminic) acids. *J. Histochem. Cytochem.* **12**, 306.

Reye, R.D.K., Morgan, G., and Barral, J. (1963). Encephalopathy and fatty degeneration of the viscera. A disease entity in childhood. *Lancet* ii, 749–52.

Roberts, G.P. (1977). Histochemical detection of sialic acid residues using periodate oxidation. *Histochem. J.* **9**, 97–102.

Roussouw, D.J., Chase, C.C., Rath, I., and Engelbrecht, F.M. (1976). The histochemical localisation of cholesterol in formalin fixed and fresh frozen sections. *Stain Technol.* **51**, 143–5.

Schultz, A. (1924). Eine methode des mikrochemischen Cholesterin-nachweises am Gewebsschnitt. *Zbl. allg. Path. Anat.* **35**, 314–17.

Schwerer, B., Lassman, H. and Bernheimer, H. (1982). Antisera against ganglioside GM_2: immunochemical and immunohistological studies. *Neuropath. appl. Neurobiol.* **8**, 217–26.

Smith, J.L. (1908). On the simultaneous staining of neutral fat and fatty acids by oxazine dyes. *J. Path. Bact.* **12**, 1–4.

Spicer, S.S. (1965). Diamine methods for differentiating mucosubstances histochemically. *J. Histochem. Cytochem.* **13**, 211–34.

Index